EURO Advanced Tutorials on Operational Research

Series editors

M. Grazia Speranza, Brescia, Italy
José Fernando Oliveira, Porto, Portugal

The EURO Advanced Tutorials on Operational Research are a series of short books devoted to an advanced topic—a topic that is not treated in depth in available textbooks. The series covers comprehensively all aspects of Operations Research. The scope of a Tutorial is to provide an understanding of an advanced topic to young researchers, such as Ph.D. students or Post-docs, but also to senior researchers and practitioners. Tutorials may be used as textbooks in graduate courses.

More information about this series at http://www.springer.com/series/13840

Lima Zhao · Arnd Huchzermeier

Supply Chain Finance

Integrating Operations and Finance in Global Supply Chains

Springer

Lima Zhao
Associate Professor of Supply Chain Management
Ningbo Supply Chain Innovation Institute China
MIT Global SCALE Network
Ningbo
China

Arnd Huchzermeier
Chair of Production Management
WHU - Otto Beisheim School of Management
Vallendar
Germany

ISSN 2364-687X ISSN 2364-6888 (electronic)
EURO Advanced Tutorials on Operational Research
ISBN 978-3-030-09549-9 ISBN 978-3-319-76663-8 (eBook)
https://doi.org/10.1007/978-3-319-76663-8

Softcover re-print of the Hardcover 1st edition 2018

Printed on acid-free paper

This Springer imprint is published by the registered company Springer International Publishing AG part of Springer Nature
The registered company address is: Gewerbestrasse 11, 6330 Cham, Switzerland

Endorsements

"This book provides a comprehensive description of the interactions between finance and operations in the management of firms and of how managers can best make decisions in recognition of these effects. Global managers and researchers alike can find valuable insights and directions for further investigation."

—John R. Birge, *Jerry W. and Carol L. Levin Distinguished Service Professor of Operations Management at University of Chicago Booth School of Business, USA*

"Supply chain finance is an emerging area where innovations are occurring that can unlock great values to complement the advances in information and physical flows of supply chain. This book is a tremendous contribution, as it provides a framework of the exciting field, as well as introducing academics and practitioners to these innovations and research opportunities."

—Hau L. Lee, *The Thoma Professor of Operations, Information and Technology at Stanford University Graduate School of Business, USA*

"This book provides an excellent overview of the field of supply chain finance and its most recent advances. Students, practitioners, and academics alike will appreciate the wealth of knowledge the authors provide in this book."

—Jan A. Van Mieghem, *Harold L. Stuart Distinguished Professor of Managerial Economics and Professor of Operations Management at Kellogg School of Management Northwestern University, USA*

"This book is indispensable for advanced students as well as practitioners when looking for a pedagogical sound and scientific rigorous approach to Supply Chain Finance."

—Ralf W. Seifert, *Professor of Operations Management at IMD Business School and Chair of Technology & Operations Management at Swiss Federal Institute of Technology in Lausanne (EPFL), Switzerland*

"This book presents a framework for integrated risk management and describes innovative approaches to risk management from a supply chain perspective. The book advances our knowledge on the interface between operations and finance and provides managerial guidelines for effective risk management in the supply chain."

—Xiande Zhao, *Professor and Philips Chair of Operations and Supply Chain Management at China Europe International Business School (CEIBS), China*

Preface

Operations and finance are two sides of the same coin. Operations management aims to match supply with demand of material flows, whereas corporate finance seeks to match supply with demand of monetary flows. Operations management sets the backbone of financial performance, and corporate finance supports real investment in operations. Integration of operations and finance leads to a sustainable competitive advantage in firms' core competencies, see Fig. 1. On one hand, real investment in operations can transform capital into material to meet customer demand. On the other hand, revenue management converts product demand into cash flow to achieve operational effectiveness (Zhao and Huchzermeier 2015). Therefore, firms can jointly optimize operations and finance across functional units in an enterprise and across supply chain partners. Operations–finance interface including supply chain finance has received growing attentions from both academia and industry in the past decades. Research on operations–finance interface is in great demand owing to its effectiveness and significance in value creation and risk management.

The structure of this book is organized as follows (as shown in Figure 1): Chapter 1 introduces concepts of operations–finance interface and a framework of integrated risk management. Chapter 2 addresses the link between capital structure and financial risk management. Chapter 3 presents concepts and techniques of operational hedging in supply chain risk management. Chapters 4 and 5 synthesize the conceptual framework and analytical modeling of integrated risk management (IRM), respectively. Chapter 6 focuses on conceptual framing and applications of supply chain finance (SCF). Chapter 7 provides a stylized model of pre-shipment finance in supply chain. Chapter 8 summarizes current landscape of operations–finance interface (OFI) and proposes directions for future explorations via an extensive research survey with bibliometric analysis.

Our book aims to provide foundations and navigation of operations–finance interface including supply chain finance primarily for graduate students (Master, Ph.D.), advanced undergraduate students, and practitioners who are interested in the

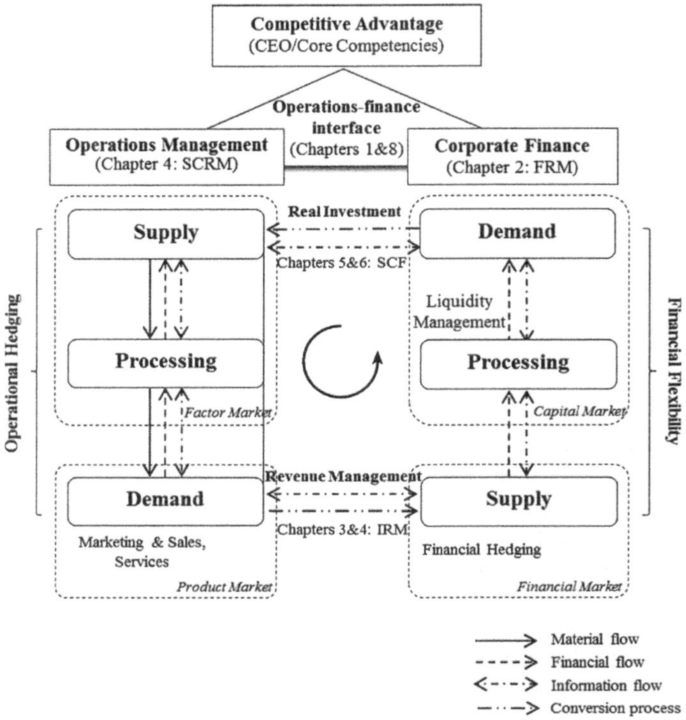

Fig. 1 Operations–finance interface and chapter overview (Adapted from Zhao and Huchzermeier 2015)

interactions between operations management and corporate finance in global supply chains. The materials in this book are presented as tutorial, theories, and models are accompanied by mini-cases and numerical examples.

Vallendar, Germany

Lima Zhao
Arnd Huchzermeier

Reference

Zhao, L., A. Huchzermeier. 2015. Operations-finance interface models: A literature review and framework. European Journal of Operational Research, 244, 905–917.

Contents

About the Authors

Lima Zhao is Associate Professor of Supply Chain Management at Ningbo Supply Chain Innovation Institute China, MIT Global SCALE Network. He was Assistant Professor of Supply Chain Finance at WHU - Otto Beisheim School of Management, Germany.

Arnd Huchzermeier (Wharton Ph.D. 1991) is Chair of Production Management at WHU - Otto Beisheim School of Management, Germany. He co-directs the INSEAD-WHU Industrial Excellence Award.

Chapter 1
Introduction to Operations–Finance Interface

This chapter presents concepts of OFI (operations–finance interface) and a framework of integrated risk management. First, we introduce a "closed-loop" view of resources by a cycle of material, financial, and information flows. Second, we synthesize various definitions on operations–finance interface to ensure consistency. Third, we present a risk management framework for multidimensional integration of operations–finance interfaces. Fourth, ten aspects are examined to specify the conditions under which firms should integrate operations and finance. Fifth, we examine the decision criteria for relationship analysis of operations and finance (complements or substitutes).

1.1 A Closed-Loop View

Operations and finance are two sides of the same coin. Operations management sets the backbone of financial performance, and corporate finance supports real investment in operations. Explorations of the relationship between operations and finance began with the Modigliani and Miller (1958) theorem, which states the *separation property* between operations and finance in perfect capital markets. According to this theorem, financial hedging can enhance firm value only by reducing (i) taxes (ii) contracting costs, or by (iii) affecting real investments in operations (Smith and Stulz 1985). Moreover, there might be value creation by (iv) financial hedging of operational risk (Gaur and Seshadri 2005).

In this chapter, we introduce a *closed-loop view* (see Fig. 1.1) to highlight operations–finance interfaces in a cycle of material, financial, and information flows. These three types of resource flows can be called the Three Bs: Boxes,

This chapter is a revised version of "Operations–finance interface models: A literature review and framework" published in issue 244 of *European Journal of Operations Research*.

L. Zhao and A. Huchzermeier, *Supply Chain Finance*, EURO Advanced Tutorials on Operational Research, https://doi.org/10.1007/978-3-319-76663-8_1

bucks, and bytes. Operations management matches supply with demand of material flows to optimize profit (Cachon and Terwiesch 2013). Meanwhile, corporate finance enables a better alignment between supply and demand of monetary flows to fund value-enhancing investments (Froot et al. 1994). In addition, managing information flows can improve the supply–demand matching of material and financial flows. The two supply–demand matching processes are connected by real investment and revenue management as a "closed-loop". *Real investment* in operations (e.g., infrastructure, human resources, technology, R&D, procurement, marketing, sales and services) can transform capital into material to meet customer demand and thereby enhance a firm's strategic positioning; *revenue management* converts product demand into cash flow to achieve operational effectiveness (Porter 1985; 1996). In practice, various sources of uncertainty may lead to supply–demand mismatches in this closed-loop. In order to generate a "virtuous" operations–finance cycle (marked in Fig. 1.1 by counterclockwise arrow), firms can adopt both operational hedging and financial flexibility to match supply with demand. In other words, to gain a sustainable competitive advantage in core competencies, firms integrate operations and finance in risk management and thereby boost the "metabolism" of the closed-loop.

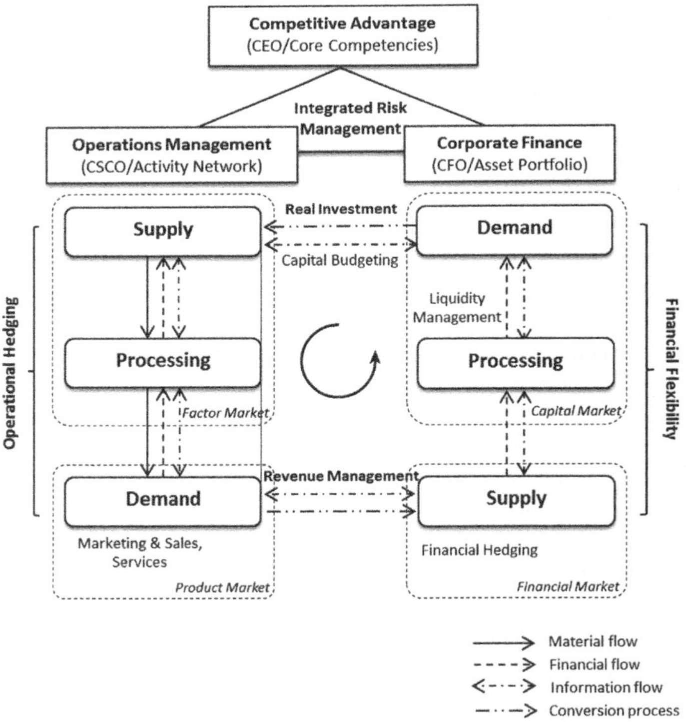

Fig. 1.1 Closed-loop view of operations and finance

Several related concepts coexist within the purview of integrated operations–finance risk management. Work on the OFI consists of "research that shows conditions under which a tighter integration of both functions within and across enterprises leads to higher value creation, while at the same time advances the knowledge and creates tools for enterprise risk management" (Birge et al. 2007), or of "integrating operations and finance in assuring profitability in the networked environment" (Kleindorfer 2012). In this book, we define the *operations–finance interface* as involving research and practice that manages operational and financial risks by joint optimization of material and monetary flows in and across enterprises.

An important subset of the OFI, *supply chain finance*, consists of "the inter-company optimization of financing as well as the integration of financing processes with customers, suppliers, and service providers in order to increase the value of all participating companies" (Pfohl and Gomm 2009), or of "the use of financial instruments, practices, and technologies to optimize the management of working capital, liquidity, and risk tied up in supply chain processes for collaborating business partners" (EBA 2014). Thus the operations–finance interface examines the joint optimization of operations and finance across both functions in an enterprise—and across organizations—within a supply chain, whereas supply chain finance focuses on the management of financial processes via collaboration of operational and financial partners within a supply chain.

Operational flexibility typically adapts the type, timing, and quantity of the material flows in an activity network contingent on realizations of uncertainty. *Operational hedging* is defined as "real (compound) options that are exercised in response to demand, price and exchange rate contingencies faced by firms in a global supply chain context" (Huchzermeier 1991), where *real options* consist of the ability to "defer, expand, contract, abandon, or otherwise alter a project at different stages during its useful operating life" (Trigeorgis 1993). Another definition of operational hedging is "mitigating risk by counterbalancing actions in the processing network that do not involve financial instruments…may include various types of processing flexibility" (van Mieghem 2003). In this book, we use the terms operational hedging and operational flexibility interchangeably.

In contrast, *financial flexibility* is defined as "the ability of a firm to access and restructure its financing at a low cost" (Gamba and Triantis 2008). Financial flexibility can transfer monetary flows across time, markets, and organizations from voluntary to binding positions through various financial instruments. A *financial instrument* is "a contract that gives rise to a financial asset of one entity and a financial liability or equity instrument of another entity" (IFRS 2012). Financial flexibility also incorporates recapitalization to match supply with demand of monetary flows and financial hedging of cash flow volatility, where *financial hedging* is "hedging that uses counterbalancing positions in financial derivative instruments" (van Mieghem 2003) and a financial derivative is "an instrument whose value depends on, or is derived from, the value of another asset" (Hull 2012).

The closed-loop view of operations–finance interfaces can be interpreted from two perspectives. On the one hand, resources flow across operational and financial units in an enterprise; on the other hand, resources flow across operational and financial partners in a supply chain. The former view leads to *enterprise risk management* (ERM), or "the ongoing proactive process of adopting a holistic approach across the enterprise to all the uncertainty which may affect either positively or negatively the achievement of its key purposes and objectives, leading to action to achieve greater business robustness and flexibility, efficient risk taking and an appropriate risk-reward balance" (ICE/FIA 2009). In contrast, the latter view results in *supply chain risk management* (SCRM), i.e., "the management of supply chain risks through coordination or collaboration among the supply chain partners so as to ensure profitability and continuity" (Tang 2006). A closely related notion, *integrated risk management*, is defined as "a multidisciplinary approach that exploits recent development in finance, decision theory, operations research, and supply chain management to manage the complex, highly interacting, and diverse global supply chain risks" (Kouvelis et al. 2012). In this book, we define *integrated risk management* as "the joint analysis, synthesis, and optimization of operational and financial risk management across functional units in an enterprise and across supply chain partners". Here, *integration* refers to (i) the joint identification/analysis of operational and financial risks; (ii) the synthesis of operations management and corporate finance (cf. Meulbroek 2002); and (iii) the unification of value-based management (flexibility/growth) and risk management (hedging/mitigation).

Traditionally, integrated risk management is the domain of a firm's CEO and CFO (Buehler et al. 2008). However, a recent SCM World survey (Lee et al. 2012) reports that chief supply chain officer (CSCO) positions have been created and have become an integral part of corporate risk management; which can lead to a C-level trilateral interaction (see Fig. 1.1) for integrated operations–finance optimization to enhance the firm's competitive advantage.

Over the last 20 years, a growing body of literature has focused on operations–finance interfaces. This research proposes conditions under which firms should integrate both functions within an enterprise and across a supply chain—which leads to ERM and SCRM, respectively—to create higher value; it also proposes ways to jointly optimize operational hedging and financial flexibility under intricate risk exposures (cf. Birge et al. 2007).

This chapter aims at providing an overview of operations–finance interface by addressing the following questions:

(i) When should firms adopt integrated operations–finance risk management?
(ii) Which steps should be taken when integrating operational hedging and financial flexibility?
(iii) How to determine whether operations and finance are complements or substitutes?

First, we introduce a "closed-loop" view of resources by a cycle of material, financial, and information flows. Second, we synthesize various definitions on operations–finance interface to ensure consistency. Third, we present a risk management framework for multidimensional integration of operations–finance interfaces. Fourth, ten aspects are examined to specify the conditions under which firms should integrate operations and finance. Fifth, we examine the decision criteria for relationship analysis of operations and finance (complements or substitutes).

1.2 Integrated Risk Management: A Framework

In order to ensure the accessibility for both academic researchers and industry practitioners, we present a tailored framework for integrated risk management based on the generic approach of ISO 31000 (ISO 2008) in Fig. 1.2. *Input* to integrated risk management includes clear definitions of risk management objectives, where various performance measures—for example, organic sales growth, adjusted earnings before interest and taxes (EBIT) margin, and adjusted earnings per share (EPS) growth—can be employed. Toward that end, risk data is collected and risk information is continuously updated with respect to the enterprise and supply chain context. Integrated risk management follows seven steps: (1) Identify risks and their interdependence, (2) specify integration conditions, (3) select operational hedging and financial flexibility, and (4) integrated optimization with relationship analysis (complements/substitutes). We shall present detailed elaborations of steps (1–4) in Sects. 1.3–1.6 to systematically review operations–finance interfaces.

Next, (5) *implementation* aims to cascade integrated risk management framework in firms' strategies, tactics, and operations. (6) *Monitoring and review* specifies and measures responsibilities and performance metrics for integrated risk management at the corporate, business unit, and functional levels. (7) *Continuous improvement* takes a top-down and bottom-up iterative approach to ensure the consistency and effectiveness of risk management steps through verification,

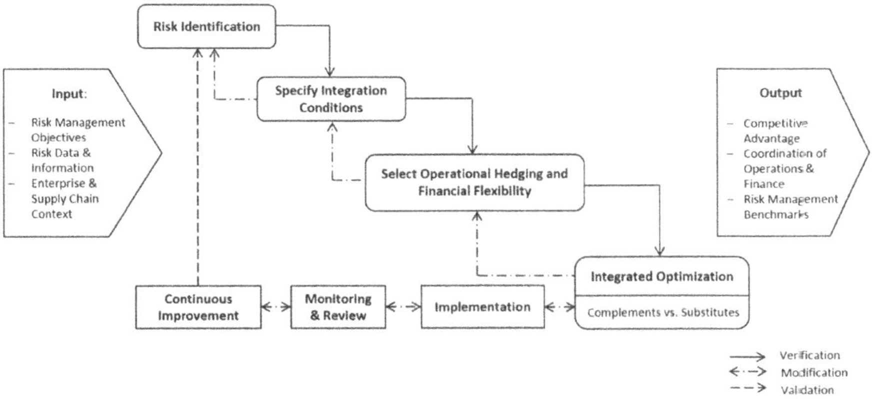

Fig. 1.2 A framework for integrated risk management

modification, and validation (marked in Fig. 1.2 by three types of arrows). *Verification* adjusts the realized risk management process to assure consistency with pre-specified objectives; *modification* enhances a previous risk management step with respect to feedback from the subsequent one; and *validation* ensures the effectiveness of risk management steps in response to input updates. The above three procedures connect the seven steps of integrated risk management as a "waterfall" model with continuous improvement. The target *output* of integrated risk management incorporates the firm's enhanced competitive advantage in its core competencies, coordination between operations and finance, as well as established risk management benchmarks.

1.3 Identify Risks and Their Interdependence

Risk identification aims to screen, categorize, and document various sources of uncertainty. *Risk* can be defined as a function of source, probability, consequence, timing, and vulnerability of an undesirable event that may occur to an organization. Closely related, *uncertainty* is "the unpredictability of environmental or organizational variables" (Miller 1992); *ambiguity* refers to "the situation where randomness cannot be expressed in terms of exact probabilities" (Natarajan et al. 2012); while *hazard* is "potential source of danger" (van Mieghem 2012). Risk incorporates both upside and downside possibilities; while uncertainty or ambiguity emphasizes on a shortfall of knowledge or information on possible outcomes; and hazard refers to the downside risk. A risk matrix is frequently used for risk analysis, which is typically a table that has several categories of "probability," "likelihood," or "frequency" for its rows and several categories of "severity," "impact," or "consequences" for its columns (Cox 2008). Risk exposure can be interpreted as the product of probability and consequence of risk.

In this book, we concentrate on two major types of risk: Operational risk and financial risk (see Table 1.1). *Operational risk* refers to uncertain time, quantity, and price in supply, processing, and demand management. *Financial risk* is "the possibility that the actual outcome is likely to diverge from the expected value" (Sharpe 1985). Based on literature survey, we classify operational risk in three categories using a process view (i.e., supply, processing, and demand risks) and financial risk in two categories: *Endogenous* financial risk includes credit and liquidity risks, which are uncertainties in corporate financing due to market imperfections. *Exogenous* financial risk refers to the uncertainties arising from volatile prices in financial market. This categorization also amplifies the interdependence between operational and financial risks, as we shall elaborate next.

Interdependence among operational and financial risks incentivizes firms to adopt integrated risk management. Operations–finance interface research has identified three types of interdependence (marked in Table 1.1 by superscripts i, ii,

Table 1.1 Risk identification

Operational risk		Financial risk	
Supply risk	Supply disruption risk[c] Uncertain supply capacity Supply yield risk Uncertain input prices Uncertain lead times	Endogenous financial risk[a]	Budget constraints Financial distress costs Bankruptcy risk Tax deductions External debt costs Agency and transaction costs
Processing risk	Processing disruption risk[c] Uncertain processing capacity Technological risk Processing yield risk Uncertain processing costs		
Demand risk	Product demand risk Market demand risk Uncertain output prices Uncertain service cost Marketing and sales risks Distribution risk	Exogenous financial risk[b]	Interest rate risk Exchange rate uncertainty[c] Asset price uncertainty Commodity price risk[c] Derivative price uncertainty

[a]Financial constraints on operations in real investment that are caused by endogenous financial risk and result in supply/processing risks
[b]Correlation between demand risk and exogenous financial risk in revenue management
[c]Alternative risk mitigation via operational hedging of financial risk and/or financial hedging of operational risk

and iii, respectively): (i) Financial constraints on operations, (ii) correlation between operational and financial risks, and (iii) alternative risk mitigation. The first two types are embedded in the two conversion processes between material and financial flows in the closed-loop (see Fig. 1.1): Real investment and revenue management. Whereas the third type lies in operational mitigation of financial risk and financial hedging of operational risk.

Financial constraints on operations refers to the bottlenecks in real investment due to market imperfections, i.e., budget constraints and costs associated with taxes, financial distress, bankruptcy, and capital. To avoid suboptimal operations (e.g., underinvestment), financial hedging and liquidity management addresses these restrictions and thereby maximizes firm value. For instance, capital constraints on newsvendor procurement can be alleviated by bank loans that achieve channel coordination (Dada and Hu 2008). Financial distress and bankruptcy risk can be jointly mitigated by liquidity management and operational hedging (Gamba and Triantis 2014). External loan financing and production technology decisions can be jointly optimized under budget constraints and a fixed bankruptcy cost (Boyabatli and Toktay 2011).

Correlation between operational and financial risks leads to a stronger link between operations and finance in revenue management. For example, the correlation between demand uncertainty and an economic or financial variable can result in joint operational and financial hedging (Mello et al. 1995). The magnitude of correlation between demand and exchange rates can determine the optimal capacity and financial hedging decisions (Chen et al. 2014). The correlated demand uncertainty with asset prices enables financial hedging of inventory risk, where the level of correlation affects the reduction in profit variance due to hedging (Gaur and Seshadri 2005). Another type of correlation can be detected between operational price risk and economic or financial variables. Price risk can be operational or financial because it may influence either type of costs. *Operational* price risk refers to uncertain input and output prices that influence operational costs, whereas *financial* price risk is the price uncertainty of assets and derivatives in financial markets. For instance, the dependence of input price on financial market movements leads to joint procurement and financial hedging (Caldentey and Haugh 2009).

Alternative risk mitigation consists of operational hedging of financial risk and financial hedging of operational risk. First, financial risk such as exchange rate uncertainty can be hedged operationally via real options and financially by currency derivatives (Ding et al. 2007). Second, operational risk that is commonly verifiable can be mitigated financially by insurance, which argues for joint insurance and operations management. For example, disruption risk can be managed by incorporating contingent supply, inventory, and business interruption insurance (Dong and Tomlin 2012). Third, price uncertainty that is both operational and financial motivates integrated risk management, e.g., commodity price risk can be managed by long- and short-term contracting and by financial hedging via commodity derivatives (Kouvelis et al. 2013).

1.4 Specify Integration Conditions: When to Integrate—Or not

Given the identified risks and their interdependence, one natural question arises: Under which conditions (i.e., prerequisites and outcomes) should firms integrate operations and finance? Owing to the heterogeneity between operational hedging and financial flexibility, operations–finance interface research has examined ten dimensions to specify integration conditions: (1) Source of uncertainty, (2) timing,

(3) availability, (4) value, (5) risk, (6) cost, (7) organizational structure, (8) supply chain structure, (9) information structure, and (10) capital structure. For ease of exposition, we classify the above ten dimensions into three categories: Feasibility (1–3), trade-offs (4–6), and structure (7–10); see Table 1.2.

To start with, firms integrate operational and financial strategies when both are *feasible* with respect to source of uncertainty, timing, and availability. *Source of uncertainty* determines the specific choices of operational hedging and financial flexibility, and the type of interdependence between operational and financial risks (see Sect. 1.3) can also affect the feasibility of a strategy—instrument portfolio. Operational strategy is more effective in mitigating operational or competitive risks, whereas financial instruments can provide a better hedge against financial and transactional risks. For instance, the presence of uncertain demand and exchange rates can lead to integrated production flexibility and financial hedging (Chowdhry and Howe 1999). Market risk can be fully hedged even in a partially complete financial market; however, demand uncertainty is private risk that can be hedged only by inventory management, not by financial instruments (Chen et al. 2007). Similar results are derived by comparing the relative effectiveness of operational and financial hedging in mitigating demand and currency risks (Ding et al. 2007).

The *timing* alignment of operational and financial decisions can also affect the feasibility of integrated risk management. Operational flexibility takes time to develop, whereas shorter maturities allow financial hedging to be implemented in a timelier manner. Thus, a financial instrument can serve as a buffer before implementing the operational strategies under risk exposure (Hommel 2003). Integrated risk management typically requires that the planning horizons of operations and finance be aligned; this can be accomplished by dynamic financial hedging in each period within long-term operational planning (Zhu and Kapuscinski 2011). The type of interdependence between operational and financial risks (see Table 1.1) can influence the relative timing of operational and financial decisions: (i) Financial constraints on operations in real investment can result in an ex ante or simultaneous financing via cash instruments or supply chain financing, such as bank loan, trade credit, and factoring (Yang and Birge 2010). (ii) Correlation of operational and financial risks in revenue management typically requires ex post or simultaneous financial hedging (Ding et al. 2007). (iii) Alternative risk mitigation can result in financial decisions that are ex ante in the case of business interruption insurance (Dong and Tomlin 2012).

Moreover, *availability* of an operational/financial strategy alters the feasible set of integrated risk management. For instance, operational hedging depends on real investments in technology choices: Product flexibility requires initial investment in dedicated or flexible production technologies (Boyabatli and Toktay 2011). Note that financial hedging may be restricted in that, for example, currency derivatives are available only for major currencies (Huchzermeier and Cohen 1996). Besides, debt financing may be subject to loan limits, such as those inherent in asset-based lending (Alan and Gaur 2012).

Firms optimize operational hedging and financial flexibility by *trading off* value, risk, and cost. The expected *value* (denoted in Table 1.2 as $\mathbb{E}[V]$) of an operational

Table 1.2 Integration conditions of operations and finance

Category	Dimension	Operational hedging	Financial flexibility	Integration conditions[a]	References
Feasibility	Source of uncertainty	More effective in operational/competitive risk mitigation	More effective to manage financial/ transactional risks	Detected interdependence[b] between operations and finance	Chowdhry and Howe (1999), Chen et al. (2007), Ding et al. (2007)
	Timing	Long-term	Short-term	Alignment of planning horizons; timing based on detected interdependence	Hommel (2003), Zhu and Kapuscinski (2011), Yang and Birge (2010)
	Availability	Determined by capacity investment	Dependent on financial market (e.g., derivatives for major currencies)	Both operational and financial hedging are feasible	Boyabatli and Toktay (2011), Huchzermeier and Cohen (1996), Alan and Gaur (2012)
Trade-offs	Value	$\mathbb{E}[V] > 0$	$\mathbb{E}[V] > 0$ (market imperfections/value max.), $\mathbb{E}[V] = 0$ (arbitrage-free)	Neither operational nor financial strategy achieves value maximization	Triantis (2000), Stulz (1996)
	Risk	Increase or decrease	Decrease	Neither operational nor financial strategy achieves risk (e.g., variance) minimization	Mello et al. (1995), Ding et al. (2007), Gamba and Triantis (2014)
	Cost	Costly, may decrease in time and volatility	Less costly, may increase in time and volatility	Integration is cost efficient	Huchzermeier and Cohen (1996), Triantis (2000)
Structure	Organizational structure	Operational unit	Financial department	Centralization benefits dominate coordination costs	Kleindorfer and Saad (2005), Glaum (2005)

(continued)

Table 1.2 (continued)

Category	Dimension	Operational hedging	Financial flexibility	Integration conditions[a]	References
	Supply chain structure	Competitive positions, (de)centralization, supply chain robustness and resilience	Decentralized hedging, competitive positions	Operational and financial is available to supply chain partners	Froot et al. (1994), Kouvelis and Zhao (2012), Gupta and Dutta (2011), Lee (2004), Christopher and Rutherford (2004), Sheffi and Rice (2005)
	Information structure	Supply chain information, market (price) information	Market information, enterprise information	Aligned incentives; optimality depends on information structure	Lee et al. (1997), Lee (2004), Caldentey and Haugh (2006), Alan and Gaur (2012)
	Capital structure	Feasible strategy set subject to capital structure	Decreases financial distress, increases debt capacity	Optimality depends on capital structure	Mello et al. (1995), Stulz (1996), Chod and Zhou (2014), Gamba and Triantis (2014)

[a]These can be model-specific and/or case-specific
[b]See Sect. 1.3 for details

strategy is typically positive because real options can limit downside risks while exploring upside potential (Triantis 2000). The value-enhancing effects of financial hedging depend largely on market assumptions and hedging objectives. In an efficient (arbitrage-free) market, the expected value of financial hedging is zero; this is known as a "self-financing" trading strategy. Under market imperfections (e.g., budget constraints and the costs associated with taxes, financial distress, bankruptcy, and external debt), the expected value of financial hedging can be positive if there are cost-reduction effects. In addition, the objective of financial hedging is value maximizing when the aim is to avoid "lower-tail outcomes" while preserving upside potential (Stulz 1996).

Risk-averse firms optimize operational and financial strategies with respect to their *risk* mitigation effects. Global supply chain network options may exploit the cash flow volatility (Huchzermeier and Cohen 1996); while allocation option can decrease the unit profit variance in global production (Ding et al. 2007). The variance effects of production flexibility reflect the extent of correlation of output price and exchange rate; thus, positive (negative) correlation leads to decreased (increased) variance (Mello et al. 1995). The variance reduction effects of financial hedging depend on the firm's objective. In an efficient market, financial hedging can be used to minimize volatility. In the presence of market imperfections, the objective of hedging can be either to maximize value or to minimize variance (Gamba and Triantis 2014).

In addition, *cost* of a risk management strategy varies as a function of availability, time, and volatility. Operational flexibility can be more costly because it requires an initial investment. For example, production flexibility depends on investments in multicountry facilities, and multisourcing flexibility depends on switching costs and development of the supplier base. In contrast, financial hedging is less costly because of lower or insignificant transaction costs. One must bear in mind that the costs of operational and financial hedging can evolve over time. Depending on the type of options and the type of hedging, the cost of real options may decrease over time (e.g., the switching costs can be spread across additional periods), whereas the cost of financial hedging may increase as the time horizon extends (Huchzermeier and Cohen 1996). Finally, greater volatility may result in more cost-effective operational flexibility and increase the cost of financial hedging; the reason is that the cost of production capacity is less sensitive to volatility than is the cost of financial derivatives (Triantis 2000).

The third category in Table 1.2 illustrates the mutual adjustments between integrated risk management and various structures: organizational, supply chain, information, and capital. On the one hand, the integrated operations–finance risk management can be tailored to these specific structures; on the other hand, firms can adapt a given structure to ensure the effectiveness of integrated risk management. *Organizational structure* in an enterprise plays an important role in mitigating risk, because the various risk exposures must be addressed by cross-functional centralization, collaboration, and coordination (the Three Cs) in ERM. Besides balancing the interactions between operations and finance, integration also leads to synergistic benefits—for example, economies of scale can arise from a firm's (cross-functional) overall hedging positions to avoid costly non-cooperative strategies (Kleindorfer and Saad 2005). However, costs may stem from the coordination among local or functional units with regard to legal and operational restrictions (Glaum 2005). Therefore, a firm's optimal structure depends on a trade-off between the benefits of centralization and the costs of coordination.

Supply chain structure affects firms' competitive and collaborative positions, and thereby determines optimal integrated risk management. Firms need not mimic the risk mitigation strategies of their rivals. If a competitor's weakness (due, e.g., to insufficient hedging) reduces the likelihood of industry overcapacity, then the benefits of investing in foreign countries may increase; this, in turn, incentivizes the firm to hedge more via currency derivatives (Froot et al. 1994). The Three Cs across supply chain partners are also required for successful SCRM. For instance, a decentralized supply chain with optimal financing and contracting under capital constraints can be coordinated via trade credit and bank financing (Kouvelis and Zhao 2012). Supply chain structure determines a firm's optimal payments to upstream and downstream partners (Gupta and Dutta 2011). At the same time, firms can adjust supply chain structure to meet market shifts and modify supply network to strategies, products, and technologies (Lee 2004). Supply chain adaptations with focus on quality control and risk management lead to robust and resilient supply chains, respectively (Christopher and Rutherford 2004). Firms can improve supply chain resilience (Sheffi and Rice 2005) by adopting operational flexibility strategies (see Chap. 3).

Information structure in the supply chain, financial markets and an enterprise can alter the optimality of risk management strategies as well. Information asymmetry among operational and financial partners in supply chain naturally affects the coordinated mitigation of risk. Information sharing may reduce the bullwhip effects and thus help to match supply with demand of material flows (Lee et al. 1997). Promoting information flow with suppliers and customers enhances supply chain agility and alignment (Lee 2004). In a financial supply chain, information asymmetry effects can be ameliorated by the design of financing mechanisms. For instance, asset-based lending can be adopted to reduce information distortion along the financial chain (Alan and Gaur 2012). We remark that a financial market's information structure can shape the optimality of joint operational and financial hedging. Complete versus partial information in financial markets yields different solutions to joint newsvendor procurement and financial hedging (Caldentey and Haugh 2006). The enterprise information structure (with respect to managers and shareholders) is a key concern of capital structure theories; that concern has spawned such notions as "separation property", "static trade-off", and "pecking order" (see Chap. 2).

Furthermore, *capital structure* determines the choice of operational and financial strategies in the face of shareholder equity concerns. For instance, the liability structure of a multinational firm shapes the interaction between production flexibility and financial hedging under agency costs of debt and bankruptcy risk. Greater flexibility reduces the number of hedging contracts needed for a given debt level but increases that number if the leverage ratio is fixed (Mello et al. 1995). Since risk management can be viewed as a substitute for equity capital, it increases the firm's debt capacity under financial distress (Stulz 1996). A firm's product flexibility can affect its optimal capital structure and thereby mitigate underinvestment, shareholder–debtholder agency conflict, and default risk (Chod and Zhou 2014). The choice of capital structure is affected by whether the objective is to maximize firm value or equity value—as is the optimality of operations, hedging, and liquidity management (Gamba and Triantis 2014).

1.5 Select Operational Hedging and Financial Flexibility

Once the integration conditions are specified and fulfilled, firms can proceed to select their portfolios of operational and financial strategies. In order to provide an overview of risk management strategies for this portfolio selection, we classify financial flexibility and operational hedging into 3 and 6 categories, respectively.

Distinguishing by valuation mechanisms, we present three categories of financial flexibility instruments: (1) A *cash instrument* is one whose value is determined directly by markets; these instruments may include bonds, loans, stocks, liquidity management, foreign currency reserves, and insurance. (2) A *supply chain instrument* is an agreement among operational and/or financial partners for transferring financial flows in a supply chain; it can take the form of a supplier subsidy, trade credit, factoring, reverse factoring, invoice discounting, or currency risk sharing. (3) A *derivative instrument* derives its price from the value of some other financial instrument or variables (e.g., futures, forwards, call and put options, swaps). Derivative markets exist for various underlying assets, including currency and interest rates, equity, credit, stocks, commodities, and even the weather. For detailed elaborations of financial flexibility instruments, please refer to Chap. 2.

Adopting a process view, we classify operational hedging strategies into six categories: (1) *Supply flexibility* incorporates multisourcing, contingent supply, backup supply, supplier improvement, and inventory mitigation to manage supply uncertainty. (2) *Processing flexibility* includes production flexibility, product flexibility, modularization, and launch flexibility to match supply and demand of material flows. (3) *Demand flexibility* uses demand shifting, allocation options, after-sales services, and entry/exit options to mitigate demand risk. (4) *Network flexibility* accounts for how supply chain network options, network configuration, coordination options, integration options, and reverse logistics are used to optimize supply chain networks. (5) *Timing flexibility*, which includes both advance flexibility and postponement flexibility, focuses on the time dimension of supply chain processes. Finally, our concept of (6) *flexibility mix* integrates the preceding five flexibilities within a category or across categories. For detailed elaborations of operational hedging strategies, please see Chap. 3.

1.6 Optimize Integrated Operations–Finance: Complements or Substitutes?

Based on the selection of operational strategies and financial instruments, firms jointly optimize their portfolios of real and financial assets. One natural question arises in joint optimization: Are operational hedging and financial flexibility complements or substitutes?

For ease of discussion, we define four types of relationship between operations and finance based on literature survey. A generic value function, V, denotes a firm's optimization objective. The value of integrated operations–finance risk management is $V(\boldsymbol{O},\boldsymbol{F}) = V(\boldsymbol{O}) + V(\boldsymbol{F}) + V(\text{IE})$. Here the adpoted operational hedging, $\boldsymbol{O} \in \mathbb{O}$ (i.e., feasible set of operational strategies) and financial flexibility, $\boldsymbol{F} \in \mathbb{F}$ (i.e., feasible set of financial instruments). $V(\text{IE})$ refers to the interaction effects between operations and finance: (i) If $V(\boldsymbol{O},\boldsymbol{F}) > V(\boldsymbol{O}) + V(\boldsymbol{F})$, i.e., $V(\text{IE}) > 0$, operational hedging and financial flexibility are complements. (ii) If $V(\boldsymbol{O},\boldsymbol{F}) = V(\boldsymbol{O}) + V(\boldsymbol{F})$, i.e., $V(\text{IE}) = 0$, operational strategy and financial instrument are separate. Moreover, when $V(\boldsymbol{O},\boldsymbol{F}) < V(\boldsymbol{O}) + V(\boldsymbol{F})$, i.e., $V(\text{IE}) < 0$, operational hedging and financial flexibility are substitutes. Two cases are present in this setting: (iii) If $V(\boldsymbol{O},\boldsymbol{F}) > \max\{V(\boldsymbol{O}), V(\boldsymbol{F})\}$, operational strategy and financial instrument are *partial* substitutes; (iv) if $V(\boldsymbol{O},\boldsymbol{F}) \leq \max\{V(\boldsymbol{O}), V(\boldsymbol{F})\}$, operational hedging and financial flexibility are *perfect* substitutes. In case (iii), the value of integration is less than the sum of two mechanisms, yet integration can still add value to the firm because its value is strictly larger than the value of any single strategy. In case (iv), either operational hedging or financial flexibility can achieve the first-best solution in the firm's objective, i.e., if the value increment of integration (defined as $\Delta V = V(\boldsymbol{O},\boldsymbol{F}) - \max\{V(\boldsymbol{O}), V(\boldsymbol{F})\}$) is negative or zero, operational strategy and financial instrument are redundant and mutually exclusive.

Global supply chain management has focused on the relationship of operational flexibility and financial hedging. Relationships among various operational and financial strategies can be analyzed by examining the marginal value increment of an additional strategy, though conclusions will differ according to model assumptions and variable changes. Table 1.3 presents an overview of analytical explorations in relationship analysis.

Production flexibility by shifting production in different countries can be adopted as a "value driver", while currency hedging via financial derivatives can tailor the variance (Hommel 2003; Ding et al. 2007). Although production flexibility and currency hedging are typically found to be complementary, they may exhibit substitution effects if operational hedging increases expected profit and reduces variance (Mello et al. 1995; Ding et al. 2007). Provided the integration of both mechanisms can lead to a positive value increment in the firm's objective, they are partial substitutes and should be adopted simultaneously and optimized jointly (Chen et al. 2014). Whereas both production and financial hedging are effective in managing currency risk, operational hedging is typically more effective in mitigating demand risk. It has been shown that financial hedging is more efficient when

Table 1.3 Overview of relationship analysis

Interdependence[a]			Operational strategy	Financial instrument	Compl.	Separ.	Partial subst.	Perfect subst.	Reference
Correlation	Constraint	Alternative							
✓	✓	✓	Production flexibility, real investment	Currency derivatives	✓				Froot et al. (1993)
✓	✓	✓	Production flexibility	Currency bonds, swaps, forwards	✓		✓		Mello et al. (1995)
✓		✓	Production flexibility	Currency forwards, options	✓		✓		Chowdhry and Howe (1999)
	✓	✓	Production diversification and flexibility	Currency forwards, options	✓		✓		Hommel (2003)
✓		✓	Capacity plan, allocation option	Currency forwards, options	✓				Ding et al. (2007)
✓			Product/postponement flexibility	Weather derivatives	✓		✓	✓	Chod et al. (2010)
	✓		Capacity ordering, inventory	Supplier subsidy	✓	✓	✓		Babich (2010)
		✓	Capacity, production, and allocation	Bonds, currency derivatives	✓	✓			Zhu and Kapuscinski (2011)
	✓		Product flexibility	Technology-specific loans	✓				Boyabatli and Toktay (2011)
		✓	Inventory, emergency sourcing	Business interruption insurance	✓		✓	✓	Dong and Tomlin (2012)
✓		✓	Capacity, outsourcing, postponement	Currency derivatives					Chen et al. (2014)
✓	✓		Production entry/exit	Cash, swaps	✓		✓		Gamba and Triantis (2014)

[a]See Sect. 1.3 for details

the firm is strongly averse to risk than otherwise (Zhu and Kapuscinski 2011; Chen et al. 2014). Furthermore, a firm's liability structure (e.g., debt level) can be part of what determines the optimal joint operational and financial hedging under market imperfections (Mello et al. 1995).

For instance, operational flexibility and financial hedging via weather derivatives can be jointly adopted to manage weather-dependent demand risk. Product flexibility and financial hedging are complements or substitutes depending on whether there is (respectively) a positive or negative correlation between demand and weather variable; postponement flexibility and financial hedging are substitutes (Chod et al. 2010). Insurance and operational strategies (inventory and emergency sourcing) can be complements or substitutes, depending on whether their net-premium interaction dominates penalty-reduction in managing disruption risk (Dong and Tomlin 2012). Supplier subsidies and inventory can be substitutes with back orders; capacity ordering and subsidies are substitutes under proportional random yield; and the relationship between order quantities and subsidies depends on how convex the manufacturer's cost function is with upfront capacity costs (Babich 2010).

Moreover, market imperfections can affect the optimality and relative effectiveness of operational hedging and financial flexibility. Operational strategies and financial instruments are typically found to be complementary, yet they may behave as partial substitutes when being jointly optimized. Ensuring monetary supply via operational flexibility, financial hedging, and liquidity management can enhance firm value through real investments and reduce the costs of market imperfections. The optimality of joint financing and real investment depends on the risk exposure of real investments and revenues, market competition, and the hedging strategies of competitors (Froot et al. 1993), as well as on the interplay of operational diversification, capacity pooling, and loan financing (Boyabatli and Toktay 2011). Liquidity management is critical to integrated risk management motivated by market imperfections (Gamba and Triantis 2014).

References

Alan, Y., & Gaur, V. (2012). *Operational investment and capital structure under asset based lending: A one period model*. Working paper, Cornell University.

Babich, V. (2010). Independence of capacity ordering and financial subsidies to risky suppliers. *Manufacturing & Service Operations Management, 12,* 583–607.

Birge, J., Kouvelis, P., & Seppi, D. (2007). Call for papers: Special Issue of *Management Science*. Interfaces of Operations and Finance.

Boyabatli, O., & Toktay, L. B. (2011). Stochastic capacity investment and flexible vs. dedicated technology choice capacity investment in imperfect capital markets. *Management Science, 57,* 2163–2179.

Buehler, K., Freeman, A., & Humle, R. (2008). Owning the right risk. *Harvard Business Review, 86,* 102–110.

Cachon, G., & Terwiesch, C. (2013). *Matching supply with demand: An introduction to operations management*. McGraw-Hill/Irwin.

Caldentey, R., & Haugh, M. (2006). Optimal control and hedging of operations in the presence of financial markets. *Mathematical Operations Research, 31,* 285–304.

Caldentey, R., & Haugh, M. (2009). Supply contracts with financial hedging. *Operations Research, 57,* 47–65.

Chen, L., Li, S., & Wang, L. (2014). Capacity planning with financial and operational hedging in low-cost countries. *Production and Operations Management, 23,* 1495–1510.

Chen, X., Sim, M., Simchi-Levi, D., & Sun, P. (2007). Risk aversion in inventory management. *Operations Research, 55,* 828–842.

Chod, J., Rudi, N., & van Mieghem, J. A. (2010). Operational flexibility and financial hedging: Complements or substitutes? *Management Science, 56,* 1030–1045.

Chod, J., & Zhou, J. (2014). Resource flexibility and capital structure. *Management Sci., 60,* 708–729.

Chowdhry, B., & Howe, J. T. B. (1999). Corporate risk management for multinational corporations: Financial and operational hedging policies. *European Finance Review, 2,* 229–246.

Christopher, M., & Rutherford, C. (2004). Creating supply chain resilience through agile six sigma. *Critical Eye,* 24–28.

Cox, L. A., Jr. (2008). What's wrong with risk matrices? *Risk Analysis, 28,* 497–512.

Dada, M., & Hu, Q. (2008). Financing newsvendor inventory. *Operations Research Letters, 36,* 569–573.

Ding, Q., Dong, L., & Kouvelis, P. (2007). On the integration of production and financial hedging decisions in global markets. *Operations Research, 55,* 470–489.

Dong, L., & Tomlin, B. (2012). Managing disruption risk: The interplay between operations and insurance. *Management Science, 58,* 1898–1915.

Euro Banking Association [EBA]. 2014. *Supply chain finance—EBA European market guide.*

Froot, K. A., Scharfstein, D. S., & Stein, J. C. (1994). A framework for risk management. *Harvard Business Review, 72,* 91–102.

Froot, K. A., Scharfstein, D. S., & Stein, J. C. (1993). Risk management: Coordinating corporate investment and financing policies. *Journal of Finance, 48,* 1629–1658.

Gamba, A., & Triantis, A. J. (2008). The value of financial flexibility. *Journal of Finance, 63,* 2263–2296.

Gamba, A., & Triantis, A. J. (2014). Corporate risk management: Integrating liquidity, hedging, and operating policies. *Management Science, 60,* 246–264.

Gaur, V., & Seshadri, S. (2005). Hedging inventory risk through market instruments. *Manufacturing & Service Operations Management, 7,* 103–120.

Glaum, M. (2005). Foreign exchange risk management in German non-financial corporations: An empirical analysis. In M. Frenkel, U. Hommel, & M. Rudolf, (Eds.), *Risk management—Challenge and opportunity* (pp. 373–393). Berlin: Springer.

Gupta, S., & Dutta, K. (2011). Modeling of financial supply chain. *European Journal of Operational Research, 211,* 47–56.

Hommel, U. (2003). Financial versus operative hedging of currency risk. *Global Finance Journal, 14,* 1–18.

Huchzermeier, A. (1991). *Global manufacturing strategy planning under exchange rate uncertainty.* Ph.D. Thesis, The Wharton School, University of Pennsylvania.

Huchzermeier, A., & Cohen, M. A. (1996). Valuing operational flexibility under exchange rate risk. *Operations Research, 44,* 100–113.

Hull, J. C. (2012). *Options, futures, and other derivatives* (8th ed.). Boston: Pearson Education.

Institution of Civil Engineers and the Faculty and Institute of Actuaries [ICE/FIA]. (2009). ERM—A guide to implementation (Draft). Institution of Civil Engineers and the Faculty and Institute of Actuaries, London.

International Organization for Standardization [ISO]. (2008). Risk management—Principles and guidelines on implementation. Draft International Standard, ISO/DIS31000.

International Financial Reporting Standards [IFRS]. (2012). Technical summary. IAS (International Accounting Standards) 32 financial instruments: presentation. International Financial Reporting Standards.

Kleindorfer, P. (2012). Foreword. In P. Kouvelis, L. Dong, O. Boyabatli, & R. Li (Eds.), *Handbook of integrated risk management in global supply chains* (pp. xiii–xiv). New York: Wiley.

Kleindorfer, P. R., & Saad, G. H. (2005). Managing disruption risks in supply chain. *Production and Operations Management, 14,* 53–68.
Kouvelis, P., Li, R., & Qing, D. (2013). Managing storable commodity risks: The role of inventories and financial hedge. *Manufacturing & Service Operations Management, 15,* 507–521.
Kouvelis P., L. Dong, O. Boyabatli, & R. Li. (2012). Integrated risk management: A conceptual framework with research overview and applications in practice. In P. Kouvelis, L. Dong, O. Boyabatli, & R. Li (Eds.), *Handbook of integrated risk management in global supply chains* (pp. 3–12). New York: Wiley.
Kouvelis, P., & Zhao, W. (2012). Financing the newsvendor: Supplier vs. bank, and the structure of optimal trade credit contracts. *Operations Research, 60,* 566–580.
Lee, H. L. (2004). The triple-A supply chain. *Harvard Business Review,* October, 102–112.
Lee, H. L., O'Marah, K., & John, G. (2012). The chief supply chain officer report 2012. SCM World.
Lee, H. L., & Tang, C. S. (1997). Modeling the costs and benefits of delayed product differentiation. *Management Science, 43,* 40–53.
Mello, A. S., Parsons, J. E., & Triantis, A. J. (1995). An integrated model of multinational flexibility and financial hedging. *Journal of International Economics, 39,* 27–52.
Meulbroek, L. (2002). The promise and challenge of integrated risk management. *Risk Management and Insurance Review, 5,* 55–66.
Miller, K. (1992). A framework for integrated risk management in international business. *Journal of International Business Studies, 23,* 311–331.
Modigliani, F., & Miller, M. (1958). The cost of capital, corporation finance, and the theory of investment. *The American Economic Review, 48,* 261–297.
Natarajan, K., Sim, M., & Teo, C. (2012). Beyond risk: Ambiguity in supply chains. In P. Kouvelis, L. Dong, O. Boyabatli, & R. Li (Eds.), *Handbook of integrated risk management in global supply chains* (pp. 103–124). New York: Wiley.
Pfohl, H.-C., & Gomm, M. (2009). Supply chain finance—Optimizing financial flows in supply chains. *Logistics Research, 1,* 149–161.
Porter, M. E. (1985). *Competitive advantage: Creating and sustaining superior performance.* New York: The Free Press.
Porter, M. E. (1996). What is strategy? *Harvard Business Review, 74,* 61–78.
Sharpe, W. (1985). *Investments* (3rd ed.). Englewood Cliffs, NJ: Prentice-Hall.
Sheffi, Y., & Rice, J., Jr. (2005). A supply chain view of the resilient enterprise. *MIT Sloan Management Review, 47,* 41–48.
Smith, C. W., & Stulz, R. M. (1985). The determinants of firms' hedging policies. *Journal of Financial and Quantitative Analysis, 20,* 391–405.
Stulz, R. (1996). Rethinking risk management. *Journal of Applied Corporate Finance, 9,* 8–25.
Tang, C. S. (2006). Perspectives in supply chain risk management. *Internationl Journal of Production Economics, 103,* 451–488.
Triantis, A. J. (2000). Corporate risk management: Real options and financial hedging. *Journal of Applied Corporate Finance, 13,* 64–73.
Trigeorgis, L. (1993). Real options and interactions with financial flexibility. *Financial Management, 22,* 202–224.
van Mieghem, J. A. (2003). Capacity management, investment and hedging: Review and recent developments. *Manufacturing & Service Operations Management, 5,* 269–302.
van Mieghem, J. A. (2012). Risk management and operational hedging: An overview. In P. Kouvelis, L. Dong, O. Boyabatli, & R. Li (Eds.), *Handbook of integrated risk management in global supply chains* (pp. 13–49). New York: Wiley.
Yang, S. A., & Birge, J. (2010). *How inventory is (should be) financed: Trade credit in supply chains with demand uncertainty and costs of financial distress.* Working paper, University of Chicago.
Zhu, W., & Kapuscinski, R. (2011). *Optimal operational versus financial hedging for a risk-averse firm.* Working paper, University of Michigan.

Chapter 2
Capital Structure and Financial Risk Management

This chapter emphasizes the link, between capital structure and financial risk management, that is central to matching the supply with demand of monetary flows. We start by investigating how capital structure theories are applied for approach choice of operations and finance (centralization or decentralization). Next, we link approach choice and relationship analysis in the context of integrated risk management. We then categorize the types of financial flexibility and describe the various financial instruments. Finally, we study the interaction between corporate investment and financial risk management.

2.1 Approach Choice: Centralization or Decentralization?

In addition to relationship analysis between operations and finance, firms face the choice of centralized versus decentralized approach to integrated risk management. Operations–finance interface models have invoked three theories of capital structure[1]—separation property, static trade-off, and pecking order—for this approach choice.

Modigliani and Miller (1958) argue that a firm's capital structure and financial decisions are independent of the optimal investment and operational decisions under perfect capital markets. This is referred to as the *separation property* of operations and finance, which should hold in the absence of information asymmetry and incentive misalignments between shareholders and managers. Subsequent research has incorporated various market frictions and information structures, which has resulted in two competing theories of capital structure: Static trade-off theory and pecking order theory. In *static trade-off* theory proposed by Modigliani and Miller (1963), a firm optimizes its liability structure by trading off the benefits

[1]For extensive reviews of capital structure research, please see Harris and Raviv (1991), Myers (2003), and Graham and Leary (2011).

L. Zhao and A. Huchzermeier, *Supply Chain Finance*, EURO Advanced Tutorials on Operational Research, https://doi.org/10.1007/978-3-319-76663-8_2

and costs of external debt under complete information. The benefits of external debt include, for instance, the tax shield and fewer free cash flow problems; the costs of debt include the costs associated with, inter alia, financial distress, bankruptcy risk, and agency problems. In contrast, Myers and Majluf (1984) and Myers (1984) propose a *pecking order* theory that incorporates manager–shareholder information asymmetry and in which firms prefer internal to external financing because of transaction and adverse selection costs. Even though managers are supposedly better informed than shareholders, the former are assumed to act in the best interests of the latter. Given these assumptions, a firm sources its capital by the following pecking order: from cash flow (internal equity), straight debt, convertible debt (with equity option), to external equity.

The separation property between operations and finance, static trade-off theory, and pecking order theory have been applied to choose a centralized versus decentralized approach to integrated risk management. To start with, recall that *separation property* refers to the case where operational and financial strategies are adopted simultaneously but optimized separately (decentralization) with zero interaction effects (see Sect. 1.6). The separation of operations and finance is typically derived under *strong* modeling assumptions (e.g., perfect capital market). Besides, this property does not imply that the optimality of operational and financial decisions is irrelevant; in fact, optimizing operational and financial decisions separately can create significant value. For instance, a given operational strategy can lead to a determinate firm value independently of financial hedging. Meanwhile, currency hedging affects the feasible set of production strategies under bankruptcy risk and agency costs of debt (Mello et al. 1995). Optimal inventory can be independent of a firm's wealth, market position, and financial hedging in a multiperiod inventory model featuring a partially complete financial market (Smith and Nau 1995), whose incompleteness stems from private demand risk (Chen et al. 2007). Optimal capacity reservation and supplier subsidy decisions can be made separately under conditions of independent supply and demand shocks, no inventory, random capacity, and zero upfront costs (Babich 2010). Under perfect (competitive) capital market assumptions, capital constrained firms can decouple its operational and bank financing decisions (Boyabatli and Toktay 2011; Kouvelis and Zhao 2012).

Moreover, joint production and financing decisions can be optimized simultaneously as a *static trade-off* between tax benefits of debt and financial distress costs under demand uncertainty and market imperfections (Xu and Birge 2004). The trade-off between investing in an external market and a newsvendor business with asset-based lending shapes the business owner's capital structure under tax codes, bankruptcy costs, and information asymmetry (Alan and Gaur 2012).

Finally, the *pecking order* theory as applied in integrated risk management suggests that financial instruments and operational strategies are partial substitutes with different execution priorities due to various costs. For example: if production diversification and flexibility incur additional costs, then financial instruments will be preferred over operational strategies (Hommel 2003). When financing expansion of pre-IPO production capacity under a budget constraint, the firm can first consume internal assets as the cheapest capital and then turn to bank loans (Babich and

Sobel 2004). In the presence of cash holding costs, loan limits, and external financing costs; real investment may be financed by a pecking order among internal funds, credit lines, and external equity (Bolton et al. 2009). To finance inventory in a budget-constrained supply chain, a retailer first chooses internal capital, then employs the cheapest trade credit, and finally diversifies external financing between trade credit and bank loan (Yang and Birge 2010). The manufacturer can choose the less expensive financing option (bank loan versus trade credit) in a supply chain under random yield, fixed supplier costs, financial constraints, and information asymmetry (Babich et al. 2012). Furthermore, the pecking order theory holds when it is optimal for a manufacturer to borrow the smallest amount from a bank that satisfies the liquidity constraint (Li et al. 2013). Table 2.1 summarizes the applications of capital structure theory in optimizing integrated risk management.

2.2 Link Relationship Analysis with Approach Choice

By connecting the relationship analysis (Sect. 1.6) and approach choice (Sect. 2.1) of integrated risk management, we find that the interaction effects and value increment jointly determine the relationship, compatibility, and approach of integrated operations and finance. Table 2.2 depicts the link between relationship analysis and approach choice in operations–finance interface models through three causal relations: (i) The interaction effects between operational hedging and financial flexibility shape their relationship, (ii) the value increment in the firm's objective determines their compatibility, and (iii) the interaction effects and value increment can jointly set the optimal approach to integrated risk management.

Intuitively, one may expect that operational hedging and financial flexibility should be optimized in a centralized (decentralized) manner if they are complements (substitutes). However, this one-to-one correspondence does not always hold for integrated risk management. We find two counterintuitive results in Table 2.2: (i) If the interaction effects are zero (separation), operational strategies and financial instruments should be adopted simultaneously yet optimized separately (decentralization). (ii) Operational hedging and financial flexibility should be optimized in a centralized manner even if they are partial substitutes.

The managerial insights from Table 2.2 are as follows: When operational hedging and financial flexibility are complements or partial substitutes, they should be optimized by centralization, collaboration, and coordination of operational and financial departments. In Case I (resp., Case III), the Three Cs (i.e., centralization, collaboration, and coordination) between operations and finance should focus on maximizing (resp., minimizing) their interaction effects. When operational hedging and financial flexibility are separate or perfect substitutes, they should be optimized independently. Case II leads to respective optimizations of operations and finance. In Case IV, a firm's risk management should incorporate *either* operational strategies *or* financial instruments.

Table 2.1 Overview of approach choice

Interdependence[a]			Market assumptions	Information structure		Separation property	Static trade-off	Pecking order	References
Correlation	Constraint	Alternative		Symmetric	Asymmetric				
✓	✓	✓	Agency costs of debt, bankruptcy costs	✓		✓			Mello et al. (1995)
	✓	✓	Efficient market, diversification/flexibility costs	✓				✓	Hommel (2003)
	✓		Financial constraints on capacity expansion	✓				✓	Babich and Sobel (2004)
✓	✓		Perfect/imperfect market (tax, bankruptcy)	✓			✓		Xu and Birge (2004)
✓			Partially complete financial market	✓		✓			Chen et al. (2007)
✓	✓		Cash holding costs, loan limits, external financing costs	✓				✓	Bolton et al. (2009)
	✓		Independent supply and demand shocks, no inventory, random capacity, zero upfront costs	✓		✓			Babich (2010)
	✓		Loan limits, competition, transaction costs		✓			✓	Babich et al. (2012)
	✓		Asset-based lending, tax, bankruptcy		✓		✓		Alan and Guar (2012)
	✓		Perfect (competitive) capital market			✓			Kouvelis and Zhao (2012)
	✓		Perfect capital market	✓		✓			Boyabatlı and Toktay (2011)
	✓		Zero long-term debt, bankruptcy costs		✓			✓	Li et al. (2013)

[a]See Sect. 1.3 for details

Table 2.2 Link of relationship analysis and approach choice

Case	Interaction Effects	Value Increment	Relationship	Compatibility	Approach
I	$V(\text{IE}) > 0$	$\Delta V > 0$	Complements	Compatible	Centralization
II	$V(\text{IE}) = 0$	$\Delta V > 0$	Separation	Compatible	Decentralization
III	$V(\text{IE}) < 0$	$\Delta V > 0$	Partial substitutes	Compatible	Centralization
IV	$V(\text{IE}) < 0$	$\Delta V > 0$	Perfect substitutes	Incompatible	Decentralization, exclusion

After joint optimization of operations and finance, efficiency and effectiveness of integrated risk management can be evaluated by performance measures such as organic sales growth, adjusted EBIT and EPS (Rorsted and Knobel 2012). Supply chain partners may trade off financial and operational measures. For instance, increases in working capital can decrease total operational cost, increase total financial cost, and lower return on working capital investment in a two-stage supply chain (Protopappa-Sieke and Seifert 2010). Successful risk management practices can be established as performance benchmarks for future implementations. Sustained value creation from integrated operations–finance optimization leads to competitive advantage in firms' core competencies (see Fig. 1.1).

2.3 Categorization of Financial Flexibility

In this section, we divide financial flexibility instruments into three categories based on their valuation mechanisms. Table 2.3 presents the categories and instruments along with their corresponding mechanisms and trade-offs.

(1) A *cash instrument* is one whose value is determined directly by markets; such instruments include loans, stocks, liquidity management, foreign currency reserves, and insurance. A *loan* refers to the act that a creditor gives monetary value to another party in exchange for future repayment of the principal amount along with interest or other finance charges. A special form of loan, asset-based financing, is offered by banks and secured by suppliers' physical assets, such as inventories and equipment (Buzacott and Zhang 2004). A technology-specific loan features a unit financing cost that depends on the firm's choice of flexible versus dedicated technologies (Boyabatlı and Toktay 2011). A *stock* is a type of security that signifies ownership in a firm and represents a claim on part of the firm's assets and earnings. "Common" stock gives the owner the right to vote at shareholders' meetings and to receive dividends; owners of "preferred" stock seldom have voting rights yet have a higher claim on assets and earnings than that of common shareholders. That is, owners of preferred stock receive dividends before common shareholders do and also have priority in the event of bankruptcy or liquidation.

Table 2.3 Categorization of financial flexibility

Category	Financial instrument	Mechanism	Trade-offs	References
Cash instruments	Loans	A certain amount of cash borrowed at a fixed interest rate from a creditor (e.g., asset-based or technology-specific)	Operational benefits versus Interests payments	Buzacott and Zhang (2004), Boyabatlı and Toktay (2011)
	Stocks	A security that represents ownership in a firm and has claims on part of the firm's assets and earnings	Bid price versus Ask price	Caldentey and Haugh (2006)
	Liquidity management	A reserve fund that consists of cash holdings in a base currency or other highly liquid assets	Liquidity benefits versus Holding costs	Gamba and Triantis (2014)
	Foreign currency	A monetary reserve of a firm in foreign currency	Reserve benefits versus Currency risk	Chowdhry and Howe (1999)
	Insurance	An equitable transfer of the risk of a loss (e.g., business interruption) from one entity to another in exchange for payment	Insurance coverage versus Insurance premium	Dong and Tomlin (2012)
Supply chain instruments	Supplier subsidy	A buyer provides financial aid to a supplier to increase its assets or reduce its liabilities	Supplier reliability versus Subsidy costs	Babich (2010)
	Trade credit	A buyer purchases goods from a supplier with optional postponed payment (extended credit)	Early payment discount versus Postponed payment price	Kouvelis and Zhao (2012)
	Factoring	A supplier sells accounts receivable to a third party at a discount for immediate cash	Liquidity benefits versus Factoring discount	Yang and Birge (2013)
	Reverse factoring	A creditor buys accounts receivable from a supplier under an acceptance from a reputable buyer who commits to pay the creditor on the due date	Liquidity benefits versus Interest rate costs	Klapper (2006)
	Dynamic discounting	A supplier offers early payment discount at various time-dependent rates to a buyer	Liquidity benefits versus Discount rates	EBA (2014)
	Currency risk sharing	A contract that allows exchange rate gains/losses to be shared equally by a supplier and a buyer	Expected payoff versus Currency uncertainty	Kouvelis (1999)

(continued)

Table 2.3 (continued)

Category	Financial instrument	Mechanism	Trade-offs	References
Derivative instruments	Futures	A standardized agreement to buy an amount of an asset at a fixed price at a standardized delivery time in the future	Hedging benefits versus Financial uncertainty	Kouvelis (1999)
	Forwards	A customized agreement to buy an asset at a fixed price at a specific time in the future		Hommel (2003)
	Call/put options	An agreement that gives holder the right to buy/sell a certain asset by a certain date for a certain strike price		Ding et al. (2007)
	Swaps	An agreement to exchange cash flows at specified future times according to certain specified rules		Hull (2012). Gamba and Triantis (2014)

Liquidity management is usually based on a reserve fund that comprises cash holdings in a base currency or other highly liquid assets. Liquidity management is an important risk management mechanism, especially when hedging tools are limited in the extent to which they can mitigate risk exposure and/or when restructuring operations involve extraordinary costs (Gamba and Triantis 2014). In practice, the importance of liquidity management has been highlighted by experience gained during the 2007–2008 financial crisis. For instance, BMW developed and rigorously adheres to a minimum liquidity standard, under which solvency is assured at all times by maintaining a liquidity reserve and a broad diversification of refinancing sources (BMW 2015). Foreign currency reserve refers to the monetary value that a firm keeps in foreign currency in order to mitigate the effects of exchange rate fluctuations (Chowdhry and Howe 1999). *Insurance* is a contract whereby one party receives financial protection, or the promise of reimbursement against losses, from another party—usually an insurance firm. Insurance companies pool their respective clients' risks so that premiums (periodic payments) are more affordable for the insured. Specific insurance products have been developed to protect firms from operational risks. For example, "business interruption" insurance protects the firm against losses incurred when it cannot operate normally because of disruption to one of its facilities or to a supplier's facility (Dong and Tomlin 2012).

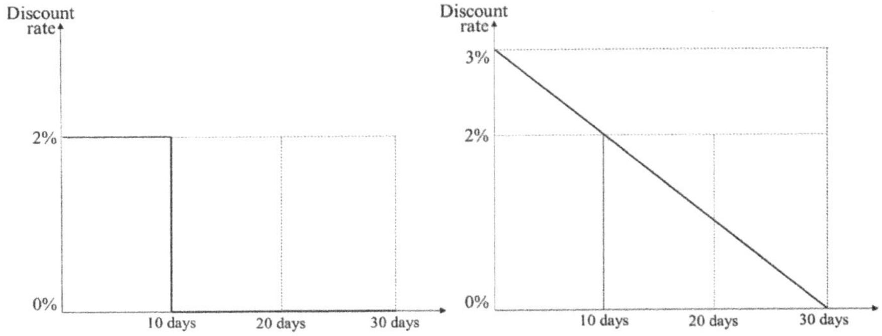

Fig. 2.1 "2/10, net 30" trade credit (left panel) and dynamic discounting (right panel)

(2) A *supply chain instrument* is an agreement among operational and/or financial partners for transferring financial flows in a supply chain. Examples include trade credit, dynamic discounting, supplier subsidy, factoring, reverse factoring, and currency risk sharing. Of these, the first two are designed to alleviate the capital constraint of a downstream buyer (retailer). *Trade credit* refers to a supply contract under which a buyer can purchase the supplier's products at a discounted wholesale price if paying early or, if paying later, at a pre-specified financing rate (Kouvelis and Zhao 2012). A frequently employed trade credit term, "2/10, net 30", means that (a) the buyer will receive 2% discount if payment is received within 10 days of the invoice date but (b) full (undiscounted) payment within 30 days is required otherwise. So under these terms, the recipient of a \$2000 invoice can either take a 2% discount (\$2000 × 0.02 = \$40) and make a payment of \$1960 within 10 days or pay the full \$2000 within the remaining 20 days. Trade credit is an example of "static" discounting—in contrast to *dynamic discounting*, which generates for suppliers the early receipt of accounts payable due from a buyer in return for a variable discount rate that declines as payment delay increases (EBA 2014). Dynamic discounting eliminates a buyer's risk of missing out on a discount if invoice approval takes longer than the 10 days cited by a "2/10, net 30" arrangement. Figure 2.1 shows that dynamic discounting allows both buyers and suppliers to set terms by placing them on a sliding scale that is open to negotiation. Thus the buyer can take the maximum allowed discount at any time within the payment window because supplier and buyer have agreed in advance on the applicable discount rates.

The following three instruments are used to alleviate the financial constraints of an upstream supplier. *Supplier subsidy* is a type of financial aid where a buyer offers to increase a supplier's assets or to reduce its liabilities (Babich 2010). For instance, when Visteon—a supplier of Ford Motor Company—considered filing for bankruptcy in 2005, the case was averted after Ford agreed to pay between \$1.6 billion

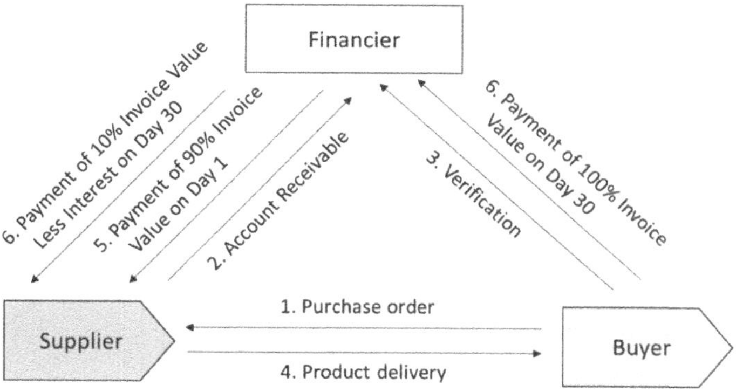

Fig. 2.2 Mechanism of a reverse factoring program

and $1.8 billion to assist Visteon's restructuring process (White 2005). Under *factoring*, a supplier sells accounts receivable to a third party (the factor) at a discount, amounting to interest plus service fees, for immediate cash. There are two types: recourse factoring and non-recourse factoring. If the factoring transfers the receivable "with recourse" then the factor has the right to collect the unpaid invoice amount from the supplier (seller); if the factoring process transfers the receivable "without recourse", then the factor must bear the loss if the buyer (account debtor) fails to pay the invoice amount. *Reverse factoring* is a financing solution initiated by the buyer (ordering party) in order to help its suppliers finance their receivables at a lower interest rate than what is offered by a financial institution. In the reverse factoring program illustrated in Fig. 2.2, the buyer has set up such a program with financier (the bank) for an onboard supplier. In this program, the financing process is as follows. First, an eligible supplier sets up an account with the bank, which is confirmed by the buyer. After product shipment, the supplier submits a disbursement request (along with a copy of the invoice) to the bank. Then bank immediately (e.g., on day 1) pays a large portion (e.g., 90%) of the invoice and subsequently pays the remaining portion (after deducting for interest) to the supplier on the agreed due date, which is typically on day 30. Meanwhile, the bank receives 100% payment from the buyer on that due date.

Currency risk sharing is an agreement whereby the gains or losses due to exchange rate fluctuations are shared equally by supplier and buyer (Kouvelis 1999). Suppose, for example, that an American firm purchases goods worth €10,000 from an European supplier and that payment is due within three months of the contract being signed. If the exchange rate at the time of the purchase contract is 1.2 (i.e., €1 = $1.20), then the payment is equivalent to $12,000 (all amounts in this example are given in USD). Now suppose that, when the three months have elapsed, the exchange rate has dropped to 1.0; in that event, the payment is worth only $10,000. Thus the European company has lost a portion of its return for no other reason than that the exchange rate has changed. Yet if a currency risk sharing

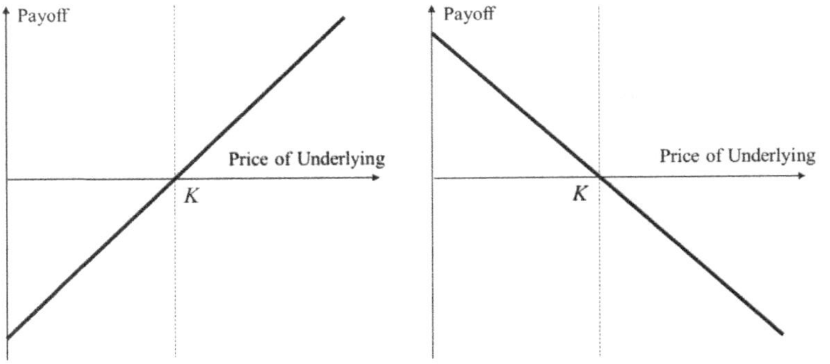

Fig. 2.3 Payoffs from a long position (left panel) and a short position (right panel) in futures/forwards

contract is in place, then that $2000 loss can be shared equally by the two parties; that is, each will absorb a $1000 loss.

(3) The price of a *derivative instrument* is based on the value of some other financial instrument or variables. Financial derivatives can be written on a variety of underlying assets: currency, interest rates, equity, credit, stocks, commodities, and even the weather. *Future* contracts are standard agreements to buy or sell an asset for a certain price at a certain time, where the *strike price* is the contract's fixed price. A long (resp. short) position in a contract refers to the obligation to buy (resp. sell) the underlying asset. Futures contracts are typically traded on an exchange, where contracts are "settled" daily. Both contractual parties in a futures contract are required to post a so-called margin (in cash or marketable securities) with the exchange clearinghouse, an action that ensures they will honor their contract commitments. *Forward* contracts are customized agreements to buy or sell an asset at a certain price at a certain future time; most such contracts are traded on the over-the-counter (OTC) market. Figure 2.3 plots the respective payoffs from long and short positions in either a future or a forward contract. For instance, if a firm holds a long position in a forward contract at a strike price of $100, then it will gain $20 if the price of the underlying asset rises to $120 but will lose $20 if that price falls to $80. Short positions exhibit an analogous (but converse) dynamic.

There are two types of *options* contract. A *call* option is a right to buy a certain asset by a certain date for a certain price (the strike price), whereas a *put* option is a right to sell a certain asset by a certain date for a certain price. Figure 2.4 illustrates the respective payoffs of call and put options. For example, if a firm purchases a call option contract with a strike price of $100 at a risk premium of $12, then it will gain $20 if the price of the underlying asset rises to $120 yet will lose only $12 (the risk premium) if that price falls to $80—because then the call option will not be executed. An analogous case can be described for a put option contract. Options contracts are traded both on exchanges and on the OTC market. An American

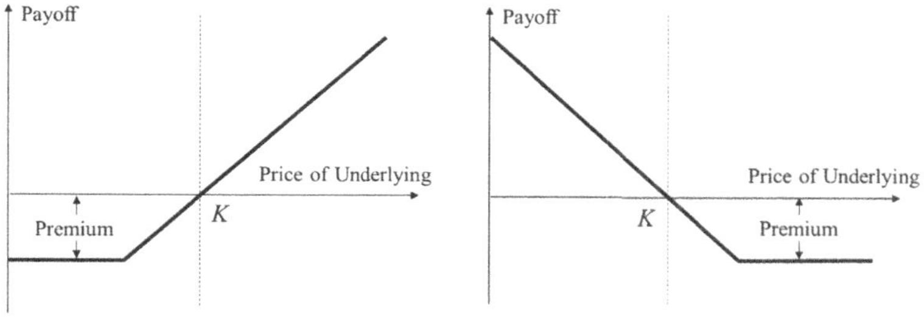

Fig. 2.4 Payoffs from call option (left panel) and put option (right panel)

option can be exercised at any time during its life, whereas a European option can be exercised only at maturity. With futures and forwards contracts, the holder has the *obligation* to buy or sell at a certain price; with options contracts, the holder the *right* to buy or sell at a certain price. Futures and forwards lock in a price for a future transaction and so are linear contracts: if a firm gains a dollar when the underlying asset moves in one direction by 10%, then it loses a dollar when that asset moves in the opposite direction by 10%. In contrast, options are nonlinear in that they allow the company to put a floor on its losses without having to give up the potential for gains. Because options can be viewed as insurance that limits downside risk without limiting upside potential, they can provide the flexibility to increase investment in good times. Finally, a *swap* is an agreement to exchange cash flows at certain future times according to certain stipulated rules (Gamba and Triantis 2014).

2.4 Link Between Corporate Investment and Financial Risk Management

Companies frequently face uncertainties due to such financial market variables as commodity prices, interest rates, and exchange rates. Globalization has naturally increased the risk exposure of international firms. In 2011, for example, the BMW Group sold only 17% (by volume) of its vehicles in Germany; emerging economies such as India, China, Russia, and Eastern Europe are now fast-growing markets. In spite of booming sales, BMW estimated that its exposure to exchange rate risk cost the firm some €2.4 billion during 2005–2009. In order to mitigate such currency risk, BMW employs two strategies. The first, "natural hedging" involves the company spending money in the markets where sales occur; this ensures that expenses and revenue are denominated in the same currency. The second strategy is financial hedging via currency derivatives, which are managed by regional treasury centers in the United States, the United Kingdom, and Singapore (Xu and Liu 2012).

According to some historians, the concept of risk management can be traced back to ancient gaming. The Old Testament relates the story of an Egyptian pharaoh who dreamed that seven healthy cattle were devoured by seven sickly cattle and that seven healthy ears of corn were devoured by seven sickly ears of corn. Joseph interpreted this dream as foretelling seven years of plenty to be followed by seven years of famine. The pharaoh hedged this risk by purchasing and storing large quantities of corn, which allowed Egypt to prosper even during the famine (Froot et al. 1994). More than two millenia ago, people played games with dice and bones. The evolution of gaming led to probability theory (as documented by Dante and Galileo), which is the basis of risk management. In the 17th century, Pascal and Fermat wrote about games of chance; their work is believed to have sparked modern probability theory.

In modern finance theory, risk is typically defined as "the possibility that actual outcomes deviate from expected ones". There are several incentives for firms to adopt risk management: (i) variability of cash flow and investment; (ii) imperfections in capital markets and underinvestment, in response to which firms can employ financial hedging to increase profits by reducing the risk of underinvesting in value-enhancing activities; (iii) costs of financial distress and bankruptcy; (iv) demands that higher payments be paid to corporate shareholders; and (v) tax benefits. We present the following examples (adapted from Froot et al. 1994) to illustrate how financial hedging can be used with regard to incentives (i) and (ii).

Example 2.1 Consider "McDermott", a multinational seller of personal care products. The firm's headquarters, research and development (R&D) department, and production plants are located in the eurozone but half of its sales revenue is from North America and Asia; as a result, the firm is exposed to exchange rate fluctuations. If exchange rates remain at their current level, then McDermott's cash flow is expected to be a million euros. When the euro depreciates significantly, revenue from non-eurozone sales yield more euros; more specifically, the firm's cash flow will increase to $a + 2b$ million euros. Conversely, if the euro appreciates significantly then its cash flow declines to $a - 2b$ million euros. We assume it is equally probable that the exchange rate will rise, fall, or stay the same.

The payoffs from McDermott's R&D programs are shown in Table 2.4. If McDermott invests a million in R&D, then a net present value (NPV) of $3c$ million can be generated. If the R&D budget is set at either $a - 2b$ million or $a + 2b$ million, then the resulting NPV will be $2c$ million. It follows that McDermott's optimal level of R&D investment is a million.

Table 2.4 McDermott's cash flows and R&D investment (millions of euros)

R&D investment	Net present value	Discounted cash flow
a + 2b	2c	a + 2b + 2c
a	3c	a + 3c
a − 2b	2c	a − 2b + 2c

Table 2.5 Value of financial hedging for McDermott (millions of euros)

Operating cash flows	R&D level w/o financial hedging	Cash flows from financial hedging	Increment in R&D investment	Value of financial hedging
a + 2b	a	−2b	0	−2b
a	a	0	0	0
a − 2b	a − 2b	2b	2b	2b + c

In the absence of financial hedging, McDermott will invest a million in R&D when the euro is stable or depreciates but can invest only $a-2b$ million when the euro appreciates. If financial markets are efficient and arbitrage-free, then hedging contracts are fairly priced; it is therefore impossible to outperform the market systematically by earning a positive expected return (cf. Chowdhry and Howe 1999; Hommel 2003). In other words, financial hedging is arbitrage-free and so the expected profit from hedging is zero. When the euro appreciates (resp. depreciates), McDermott gains (resp. loses) $2b$ million euros from financial hedging. Yet hedging results in a stable cash flow of a million, in which case McDermott can invest the optimal amount in R&D and thereby achieve an NPV of $3c$ million euros in all three scenarios. The value increment from financial hedging stems from the scenario in which the euro appreciates, because in that case the additional investment in R&D increases McDermott's NPV from $2c$ million to $3c$ million euros. Hence, the company is better-off by hedging, and the expected profit from financial hedging amounts to $\frac{1}{3}(3c-2c)=\frac{1}{3}c$ million euros. Table 2.5 summarizes the value of financial hedging for McDermott.

In this example, financial hedging aims to align the supply of funds with their demand. McDermott's cash flow supply fluctuates with the exchange rate, but its the demand for funds—that is, the optimal investment in R&D—is constant. Financial hedging transfers monetary flow from voluntary to binding positions in various scenarios; thus it reduces the variance of cash flow and thereby enables the optimal level of investment. Although the financial hedging is arbitrage-free, it may create value via investments in value-enhancing projects (cf. Smith and Stulz 1985).

Example 2.2 Now suppose that, with all other factors held equal, McDermott considers investing in a new plant in China. In this example, the exchange rate's variability affects not only McDermott's cash flows but also the value of its new plant investment. If the exchange rate remains at its current level then the cash flow of McDermott is expected to be a million euros, which is its optimal level of investment. If the euro *depreciates* significantly then that cash flow will increase to $a+2b$ million euros (because foreign sales revenue then translates into more euros) and the optimal investment will increase to $a+b$ million euros (because a new plant in China will cost more in euros). If the euro instead *appreciates* significantly, then McDermott's cash flow will fall to $a-2b$ million euros and its optimal investment to $a-b$ million euros. As before, the scenarios of depreciation, appreciation, and stability are assumed to be equally probable. In order to match the supply and

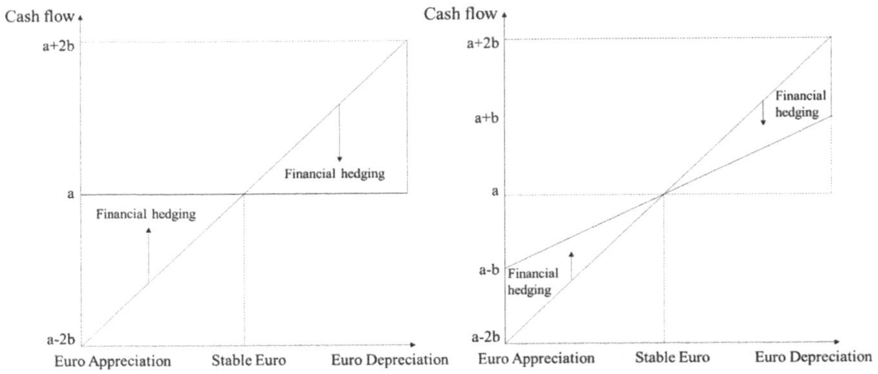

Fig. 2.5 Comparative magnitude of McDermott's hedging in Examples 2.1 and 2.2

demand of funds, McDermott employs arbitrage-free financial hedging. When the euro appreciates (resp. depreciates), the company gains (resp. loses) b million from hedging. Here McDermott need not hedge to the same extent as in the previous example because the firm has a built-in hedge with respect to foreign investment. Figure 2.5 reveals that the optimal hedge for McDermott in Example 2.2 is half of its counterpart in Example 2.1. The reason is that there is less of a mismatch between the supply and demand of funds when both cash flow and the investment opportunity fluctuate with the exchange rate.

Which is the more important goal of financial hedging: to reduce cash flow volatility or to secure an investment opportunity? In Example 2.2, if McDermott reduced its cash flow variance to zero then it would, in effect, adopt the same financial hedging strategy as in Example 2.1. However, it would then lose its assurance that it could invest the optimal amount. Thus McDermott would be b million euros short of the optimal investment amount if the euro depreciated or have b million euros more than its optimal investment level if the euro appreciated. So here, a full hedge fails to guarantee that real investment will be value-enhancing. It follows that the aim of financial hedging should be to match the supply and demand of monetary flows.

Example 2.3a Two US automotive manufacturers, Alpha and Beta, source from China and Canada (respectively) for their components. The unit purchase price is higher in Canada than in China. Supply contracts are written in the local currencies but selling prices are denominated in US dollars. In this example, which firm is more in need of a financial hedge against currency rates? If the US dollar depreciates while the Canadian dollar and the Chinese yuan rates remain stable, then both sourcing prices rise in terms of the US dollar. In this scenario, the unit purchase price of components from Canada (but not from China) will exceed the market price in US dollars. Then Beta must either cease its sales or bear losses from continued operation—an indication that its cash flow is relatively more sensitive to exchange rate fluctuations. Therefore, financial hedging is more valuable to Beta than to Alpha.

Example 2.3b Suppose now that Alpha and Beta are contemplating investments of new plants in China and Canada, respectively. Both companies are in the same respective cash flow situations as in Example 2.3a. Which firm should hedge more? If the US dollar depreciates while the Canadian dollar and the Chinese yuan remain stable, then the investment opportunities of both firms become more expensive in terms of the US dollar. Hence it may no longer be profitable to invest in Canada (since it is more expensive) even as it may still be profitable to invest in China. The US dollar's depreciation has the same effect on both firms' cash flows, yet Beta's investment opportunity would be less profitable than that of Alpha's. Thus, for Beta, the demand for funds is more aligned with their supply (i.e., there is a built-in hedge); this means that Alpha needs to hedge more than does Beta.

Comparing Examples 2.3a and 2.3b, we can see that firms operating in the same industry should not necessarily employ the same financial hedging strategy. In any case, the goal of risk management is to align the firm's cash flow supply with its demand for real investment.

Example 2.3c In this example, both Alpha and Beta are considering investments of new plants in China; all other factors in Example 2.3b remain unchanged. Suppose Beta does not adopt a financial hedge against the exchange rate; in this case, should Alpha adopt financial hedging? If Beta does not hedge, then it might be short of Chinese yuan if the US dollar depreciates. Those circumstances would reduce the likelihood of Beta investing in China, which in turn would reduce the likelihood of industry overcapacity. Hence the investment opportunity in China becomes more profitable for Alpha. We conclude that Alpha *should* hedge when Beta does not.

Example 2.3c shows that a firm should closely monitor the financial hedging strategy of its competitor(s), examine the effect of such strategy on its own investment opportunities, and then adjust its hedging strategy accordingly.

In sum: Managers can identify a firm's optimal financial hedging strategy by answering three questions, as follow. How volatile is the firm's cash flow with respect to such market variables as exchange rates, commodity prices, and interest rates? How sensitive is the optimal investment level to those financial variables? Which financial hedging strategy best matches the firm's cash flow with its optimal investment level?

Recall from the text preceding Example 2.1 that there are five incentives for adopting risk management practice. We now address the third, fourth, and fifth of those incentives in turn.

(iii) Financial hedging reduces the variance of the firm's cash flow; that, in turn, can eliminate the possibility of default. Figure 2.6 illustrates the effect of financial risk management on firm value in the presence of bankruptcy costs. The distribution of firm value V before bankruptcy *without* hedging is

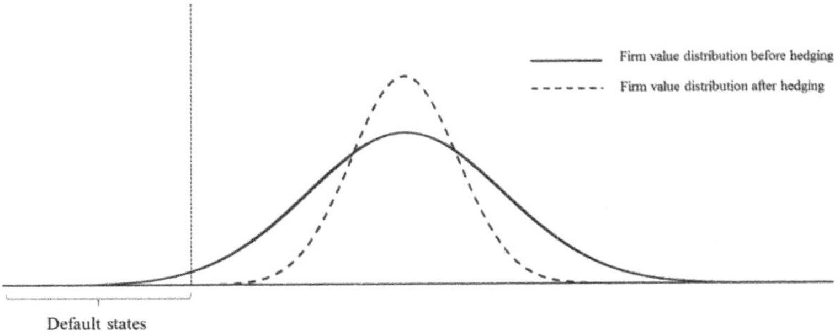

Fig. 2.6 Risk management, bankruptcy costs, and firm value

marked by the dashed curve, while the distribution of firm value *with* financial hedging has a narrower span and so does not include any default states. Thus financial hedging increases firm value by eliminating the bankruptcy cost B. By reducing the costly lower-tail outcomes (Stulz 1996), the value increment from financial risk management can be calculated as the bankruptcy costs multiplied by the probability of bankruptcy. Suppose, for instance, that the current firm value is $V = \$500$ million; if the proportional bankruptcy cost is 20% of firm value (i.e., $B = 0.2 \times V$) and if the bankruptcy probability is $P = 10\%$, then the value of financial hedging (FH) due to its ruling out bankruptcy is $V\,[\text{FH}] = B \times P = \$500$ million $\times$ 20% $\times$ 10% = \$10 million—or 2% of firm value. Hence, the value from financial hedging increases with the probability of bankruptcy (e.g., when operating cash flow declines).

(iv) A firm's shareholders or owner-managers may require a certain rate of return that reflects extant systematic and nonsystematic risks. Financial hedging can create value by reducing owners' risks and hence the required rates of return. Besides investors, other stakeholders of the firm—including managers, employees, customers, and suppliers—may require higher compensation to the extent that firm-specific value is exposed to higher risk. As a result, financial hedging can add value by reducing risk exposure and thereby enabling higher payments.

(v) The potential tax benefits from financial hedging reflect the reduced variability of income and the convexity of most tax codes. Higher reported income is typically taxed at a higher rate, whereas losses can lead to lower rates or even tax rebates. Given the convexity of tax codes, financial hedging can tailor the firm's taxable income so that it lies within the interval in which tax rates are relatively low. Hence lower variability of pre-tax income, which follows from successful financial risk management, can lead to optimal corporate tax deductions. Equivalently, when a firm's after-tax profit is concave with respect to its pre-tax earnings, then it should hedge if a reduced variability in pre-tax earnings leads to tax reduction benefits that exceed the cost of financial hedging (Smith and Stulz 1985).

References

Alan, Y., & Gaur, V. (2012). *Operational investment and capital structure under asset based lending: A one period model*. Working paper, Cornell University.

Babich, V. (2010). Independence of capacity ordering and financial subsidies to risky suppliers. *Manufacturing & Service Operations Management, 12,* 583–607.

Babich, V., Aydin, G., Brunet, P.-Y., Keppo, J., & Saigal, R. (2012). Risk, financing and the optimal number of suppliers. In H. Gurnani, A. Mehrotra, & S. Ray (Eds.), *Managing supply disruptions* (pp. 195–240). London: Springer.

Babich, V., & Sobel, M. J. (2004). Pre-IPO operational and financial decisions. *Management Science, 50,* 935–948.

BMW. (2015). BMW annual report 2014.

Bolton, P., Chen, H., & Wang, N. (2009). *A unified theory of Tobin's q, corporate investment, financing, and risk management*. Working paper, Columbia University.

Boyabatli, O., & Toktay, L. B. (2011). Stochastic capacity investment and flexible vs. dedicated technology choice capacity investment in imperfect capital markets. *Management Science, 57,* 2163–2179.

Buzacott, J., & Zhang, R. (2004). Inventory management with asset-based financing. *Management Science, 50,* 1274–1292.

Caldentey, R., & Haugh, M. (2006). Optimal control and hedging of operations in the presence of financial markets. *Mathematical Operations Research, 31,* 285–304.

Chen, X., Sim, M., Simchi-Levi, D., & Sun, P. (2007). Risk aversion in inventory management. *Operations Research, 55,* 828–842.

Chowdhry, B., & Howe, J. T. B. (1999). Corporate risk management for multinational corporations: Financial and operational hedging policies. *European Finance Review, 2,* 229–246.

Ding, Q., Dong, L., & Kouvelis, P. (2007). On the integration of production and financial hedging decisions in global markets. *Operations Research, 55,* 470–489.

Dong, L., & Tomlin, B. (2012). Managing disruption risk: The interplay between operations and insurance. *Management Sci., 58,* 1898–1915.

Euro Banking Association [EBA]. (2014). *Supply chain finance—EBA European market guide*.

Froot, K. A., Scharfstein, D. S., & Stein, J. C. (1994). A framework for risk management. *Harvard Business Review, 72,* 91–102.

Gamba, A., & Triantis, A. J. (2014). Corporate risk management: Integrating liquidity, hedging, and operating policies. *Management Science, 60,* 246–264.

Graham, J., & Leary, M. (2011). A review of empirical capital structure research and directions for the future. *Annual Review of Financial Economics, 3,* 1–37.

Harris, M., & Raviv, A. (1991). The theory of capital structure. *Journal of Finance, 46,* 297–355.

Hommel, U. (2003). Financial versus operative hedging of currency risk. *Global Finance Journal, 14,* 1–18.

Hull, J. C. (2012). *Options, futures, and other derivatives* (8th ed.). Boston: Pearson Education.

Klapper, L. (2006). The role of factoring for financing small and medium enterprises. *Journal of Banking Finance, 30,* 3111–3130.

Kouvelis, P. (1999). Global sourcing strategies under exchange rate uncertainty. In S. Tayur, R. Ganeshan, & M. Magazine (Eds.), *Quantitative models for supply chain management* (pp. 625–668). Dordrecht: Kluwer Academic Publishers.

Kouvelis, P., & Zhao, W. (2012). Financing the newsvendor: Supplier vs. bank, and the structure of optimal trade credit contracts. *Operations Research, 60,* 566–580.

Li, L., Shubik, M., & Sobel, M. (2013). Control of dividends, capital subscriptions, and physical inventories. *Management Science, 59,* 1526–5501.

Mello, A. S., Parsons, J. E., & Triantis, A. J. (1995). An integrated model of multinational flexibility and financial hedging. *Journal of International Economics, 39,* 27–52.

Modigliani, F., & Miller, M. (1958). The cost of capital, corporation finance, and the theory of investment. *The American Economic Review, 48,* 261–297.
Modigliani, F., & Miller, M. H. (1963). Corporate income taxes and the cost of capital: A correction. *The American Economic Review, 53,* 433–443.
Myers, S. C. (1984). The capital structure puzzle. *Journal of Finance, 39,* 575–592.
Myers, S. C. (2003). Financing of corporations. In G. Constantinides, M. Harris, & R. Stulz (Eds.), *Handbook of the economics of finance: Corporate finance* (pp. 215–254). Amsterdam: Elsevier.
Myers, S. C., & Majluf, N. S. (1984). Corporate financing and investment decisions when firms have information that investors do not have. *Journal of Financial Economicss, 13,* 187–221.
Protopappa-Sieke, M., & Seifert, R. W. (2010). Interrelating operational and financial performance measurements in inventory control. *European Journal of Operational Research, 204,* 439–448.
Rorsted, K., & Knobel, C. (2012). The relevance of management control for Henkel. Presentation, Henkel AG.
Smith, C. W., & Stulz, R. M. (1985). The determinants of firms' hedging policies. *Journal of Financial and Quantitative Analysis, 20,* 391–405.
Smith, J. E., & Nau, R. F. (1995). Valuing risky projects: Option pricing theory and decision analysis. *Management Science, 41,* 795–816.
Stulz, R. (1996). Rethinking risk management. *Journal of Applied Corporate Finance, 9,* 8–25.
White, J. (2005). Ford to pay up to $1.8b on Visteon. *Wall Street J.* (May 26) A3.
Xu, B., & Liu, Y. (2012). Managing foreign exchange risk: BMW moved production overseas. October 30, 2012.
Xu, X., & Birge, J. (2004). *Joint production and financing decisions: Modelling and analysis.* Working paper, Northwestern University.
Yang, S. A., & Birge, J. (2010). *How inventory is (should be) financed: Trade credit in supply chains with demand uncertainty and costs of financial distress.* Working paper, University of Chicago.
Yang, S. A., & Birge, J. (2013). *How inventory is (should be) financed: Trade credit in supply chains with demand uncertainty and costs of financial distress.* Working paper, University of Chicago.

Chapter 3
Supply Chain Risk Management

In this chapter, we introduce concepts and techniques of operational hedging in SCRM (supply chain risk management) that are used to match the supply with demand of material flows. First, we present a conceptual framework of supply chain risk management, after which we propose a classification of operational hedging based on our description of each operational strategy. We then examine representative analytical models of SCRM and give numerical examples. Finally, we discuss the empirical research on SCRM and provide some illustrative mini-cases.

3.1 Conceptual Framework of SCRM

In order to establish a framework for supply chain risk management, we begin with the concepts of risk and uncertainty. Risk is "an undesirable possible consequence of uncertainty" (van Mieghem 2012). In economics, risk refers to "situations in which we can list all possible outcomes and we know the likelihood that each outcome occurs" (Pindyck and Rubinfeld 1989). This latter definition implies that any uncertainty in outcomes, whether or not it is favorable, constitutes a risk; thus that definition covers both upside potential and downside risk. A key distinction from the common interpretation of risk is the absence of "danger" or of an "adverse event". Risk is a combination of three factors: the *probability* that an event or outcome will occur, the *consequences* of that event, and *exposure* or a causal pathway leading to the event. Risk results from uncertainty, which can be about physical, social, political, economic, cultural, environmental, or psychological events. The relationship between sources of uncertainty and risks faced by an organization can be viewed as a funnel model. Sources of uncertainty encountered by an organization can trigger two major risks that affect corporate performance—namely, operational risks and financial risks.

Supply chain risk has been defined as "the potential occurrence of an incident or failure to seize opportunities with inbound supply in which its outcomes result in a

L. Zhao and A. Huchzermeier, *Supply Chain Finance*, EURO Advanced Tutorials on Operational Research, https://doi.org/10.1007/978-3-319-76663-8_3

financial loss for the [purchasing] firm" (Zsidisin 2005), and *supply chain management* refers to "the management of upstream and downstream relationships with suppliers and customers in order to deliver superior customer value at less cost to the supply chain as a whole" (Christopher 1992). These two definitions are synchronized in the definition of *supply chain risk management* as "the management of supply chain risks through coordination or collaboration among the supply chain partners so as to ensure profitability and continuity" (Tang 2006).

Risk analysis is a term used to describe estimating the probability and consequences of uncertain events as well as an entity's exposure to risk. A *risk matrix* (see Fig. 3.1) measures sources of uncertainty in terms of their frequency (likelihood) and impact (effects). Timing analysis focuses on the time axis of risk; the effects of exposure to risk of a given magnitude may exhibit considerable variance as a function of time—for example, depending on whether a firm is at the entrepreneurial stage or at a more established stage. Vulnerability analysis characterizes the susceptibility of an organization to the adverse effects of risk. Quantitative risk analysis may involve risk matrices, timing analysis, and vulnerability analysis, the combination of which enables further evaluation of distinct risk "cells" (i.e., the distinctly demarcated areas of a risk matrix).

Risk evaluation includes the measurement and prioritization of risk cells based on risk analysis. In risk measurement, one compares the analyzed risk levels with risk criteria; the result of such measurement is an indication of which risks are entirely acceptable (the white cells in Fig. 3.1), can be tolerated (light gray cells), merit sustained attention (medium gray), or require immediate response (dark gray). Risk perception is the subjective cognitive activity of estimating the extent of risk

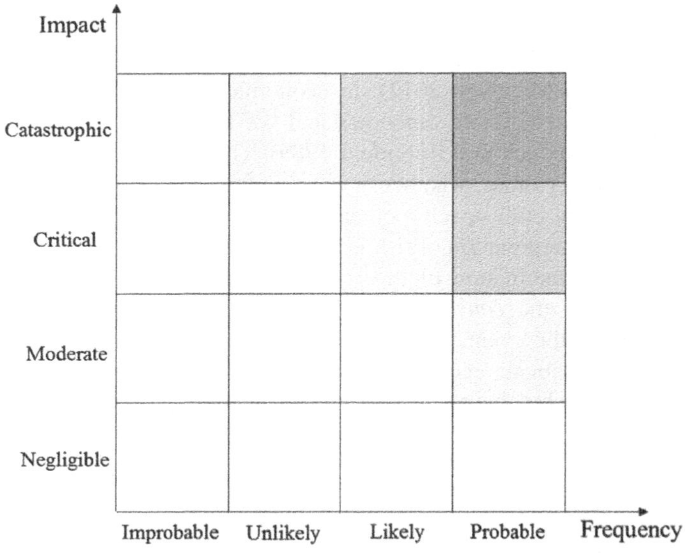

Fig. 3.1 Predictive risk matrix

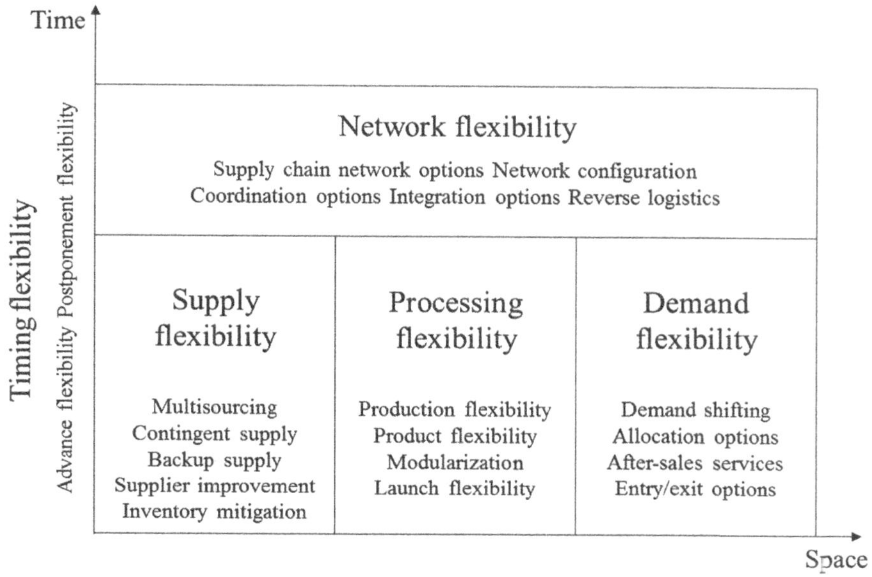

Fig. 3.2 Framework for supply chain risk management

based on interpretations of objective phenomena. Perceived risk levels can shape the risk attitudes in decision making, where the notion of "risk attitude" quantifies the extent to which decision makers are willing to adopt a course of action whose outcomes are not guaranteed (Roszkowski and Davey 2010). Risk prioritization is based on the characteristics of risk matrix cells—as determined by the risk measurement previously described—and indicates the areas on which risk management should focus (and in what order).

Figure 3.2 summarizes our proposed framework for supply chain risk management. In this framework, operational flexibility strategies are classified along the dimensions of space and time. Strategies that adjust the quality, quantity, reliability, design, or location of activities in a product's life cycle are instances of supply flexibility, processing flexibility, or demand flexibility, whose combination yields network flexibility—along the figure's horizontal (space) axis. The advancement or deferment of such flexibility is captured by timing flexibility, represented by the figure's vertical (time) axis. In addition to the five categories illustrated in Fig. 3.2, the synchronization among the operational strategies determines what we refer to as "flexibility mix" (the last category listed by Table 3.1 in Sect. 3.2).

Table 3.1 Categorization of operational hedging

Category	Operational strategy	Mechanism	Trade-offs	References
Supply flexibility	Multisourcing	Purchasing one product from multiple suppliers	Supplier diversification/ competition benefits versus Setup costs	Kouvelis (1999), Babich et al. (2007), Dada et al. (2007)
	Contingent supply	Reserving the right (capacity) to purchase a certain number of units from a supplier by quantity-flexible contracts	Contingency benefits versus Reservation costs	Tomlin (2006)
	Backup supply	Sourcing from an alternative internal (backup production) or external supply (backup supplier) in case of shortage/disruption	Marginal revenue versus Unit supply cost	Yang et al. (2009), Sting and Huchzermeier (2010)
	Supplier improvement	Investing in suppliers to improve production processes	Improvement benefit versus Investment costs	Wang et al. (2010)
	Inventory Mitigation	Holding inventory to fulfill demand under supply shortage/disruption	Expected profit versus Holding costs	Tomlin (2006)
Processing flexibility	Production flexibility	Shifting production between plants in different countries	Flexibility benefits versus Switching costs	Kazaz et al. (2005), Ding et al. (2007)
	Product flexibility	Producing multiple products from a common resource	Marginal value versus Unit capacity cost	van Mieghem (2007)
	Modularization	Assembling final product from a set of standardized components	Modularization value versus Configuration cost	Ernst and Kamrad (2000)
	Launch flexibility	Introducing new products and product varieties	Launch benefits versus R&D/ marketing costs	Vickery et al. (1999)
Demand flexibility	Demand shifting	Shifting demand across time, markets, and/or products	Flexibility value versus Shifting costs	Tang (2006)
	Allocation option	Delivering products to markets responsively and cost effectively	Expected pro fit versus Allocation costs	Ding et al. (2007)
	After-sales service option	Providing customized services and maintenance after sales	Service bene fits versus Operating costs	Kim et al. (2007)
	Entry/exit options	Expanding new market; holding back products, stopping production, or exiting market	Flexibility value versus Options costs	van Mieghem and Dada (1999), Gamba and Triantis (2014)

(continued)

Table 3.1 (continued)

Category	Operational strategy	Mechanism	Trade-offs	References
Network flexibility	Supply chain network options	Designing supply chain network by altering supply, production, and/or distribution options	Flexibility value versus Switching costs.	Huchzermeier and Cohen (1996)
	Network configuration	Configuring networks that allow for multiple supply, products, processing, and/or storage points	Configuration benefits versus Resource investment	Tomlin and Wang (2005), van Mieghem (2007)
	Coordination option	Coordinating decisions among supply chain members via contracting or information sharing	Coordination value versus Coordination cost	Cachon (2003), Chen (2003)
	Integration option	Integrating information, material, and financial flows of supply chain via merger/acquisition or IT	Integration benefits versus Integration investment	Rai et al. (2006), Nagurney (2009)
	Reverse logistics	Operating a closed-loop supply chain over the product life cycle with value recovery from returns	Return benefits versus Operating costs	Guide and Van Wassenhove (2009)
Timing flexibility	Advance flexibility	Placing orders in advance using quick response, supply contracts, and/or dual sourcing from forward and spot markets	Flexible timing value versus Implementation costs	Fisher and Raman (1996), Donohue (2000), Kouvelis et al. (2013)
	Postponement flexibility	Postponing capacity, production, and/or pricing decisions until after the realization of uncertainty		Lee (1996), van Mieghem and Dada (1999)
Flexibility mix	Inter-category mix	Combining flexibilities across various categories	Flexibility value versus Operational costs	Ernst and Kamrad (2000)
	Intra-category mix	Combining flexibilities within a certain category		Tomlin (2006)

3.2 Categorization of Operational Hedging

In this section we adopt a process view while elaborating on the operational hedging strategies pertinent to our six categories of flexibility; see Table 3.1 for an overview.

(1) *Supply flexibility* typically focuses on the decisions associated with supply quantity, timing, and reliability. *Multisourcing* refers to purchasing a given component from various suppliers—in contrast to single sourcing, under which just one supplier is used. Multisourcing diversifies risk within the supply base and may reduce prices by introducing supplier price competition. It also gives firms the flexibility to switch among suppliers and thereby lessens its dependence on any one supplier; hence the supplier relationship becomes more transactional in nature. That said, single sourcing could yield a cost advantage in that a sole supplier could reduce its production cost via the economies of scale enabled by a larger order quantity.

Single sourcing can also lower transaction costs. Furthermore, a buyer may have relatively greater access to a single supplier's know-how, which would enable faster development of new products in this more cooperative supplier relationship. A comparison of the advantages associated with single sourcing versus multi-sourcing is presented in Fig. 3.3.

Contingent supply is defined as the reservation and execution of volume flexibility, with an existing supplier, that enables temporarily increased sourcing quantities in case of future supply disruption or shortage. In 2000, for example, the semiconductor supply of Philips was disrupted owing to fire at a production plant that resulted in a shortage of cell-phone chips for both Nokia and Ericsson. Nokia was able to reroute output from the Philips Eindhoven plant and also secured increased production from alternative suppliers. Yet Ericsson's single-sourcing approach left it without a "plan B"; as a result, it incurred a loss of at least $400 million (Latour 2001; Tomlin 2006). *Backup supply* is the sourcing of a product from alternative capacity or suppliers to mitigate the effects of any supply disruption. *Supplier improvement* refers to the investment of funds and effort toward the end of improving the reliability of a supplier's production process (Sting and Huchzermeier 2010). The *inventory mitigation* strategy involves holding inventory (to fulfill demand as needed) as a means of managing supply uncertainty (Tomlin 2006).

(2) *Processing flexibility* allows for adjustment of the location, quantity, type, and design of products to match the supply of and demand for material flows. This category includes *production flexibility* (also referred to as production hedging), which shifts manufacturing quantities of a given product between plants located in different currency zones as a way to reduce exchange rate risk (Kazaz et al. 2005). In contrast, *product flexibility* refers to the production of multiple products at a single site (van Mieghem 2007). *Modularization* references product design that enables a final product's assembly based on standardized

Advantages of single sourcing	Advantages of multisourcing
Cost reduction through bundling/standardization	Lower prices due to higher competition
Smaller number of suppliers and user interfaces	Lower dependence on single suppliers
Lower transaction and management costs	Lower dependence on single technologies
Easier quality assurance and higher specialization	Flexible change of suppliers
etc.	etc.

Advantages of cooperative supplier relationship	Advantages of transactional supplier relationship
Strategic cost reduction	Lower supplier management costs
Commitment of suppliers	Price reduction potential due to higher competition
Transfer of suppliers' know-how	Higher flexibility in case of lower switching costs
Faster development of new products	Lower dependence on a single supplier
Joint planning and information sharing with suppliers	No transfer of know-how to supplier
Earlier detection of misleading developments	No decline in supplier motivation due to long-term contracts
Higher quality levels	etc.
Simpler sourcing processes	
Inventory reduction	
etc.	

Fig. 3.3 Single sourcing versus multisourcing. Adapted from Blome and Henke (2009)

components (Ernst and Kamrad 2000), and *launch flexibility* consists of the freedom to launch new products and to extend current product lines with new varieties (Vickery et al. 1999).

(3) *Demand flexibility* alters the timing, location, and type of demand for a product or service as well as strategic market choices. *Demand shifting* transfers customer demand across time, markets, and products. For instance, the advance purchase discounts offered by airlines enable customers to receive price discounts for early commitment and payment. On the one hand, customers can enjoy a price advantage by pre-booking and purchase; on the other hand, the airline benefits because pre-commitment by customers reduces its demand uncertainty. Firms can also adopt "responsive pricing" to shift the demand for one product to a substitute product (Chod and Rudi 2005). Another example of demand shifting is that of firms that produce "style goods". Thus a manufacturer of ski wear operates in two markets with selling seasons that do not overlap: the winter season runs from December to February in Europe and North America but from June to August in Oceania. The firm can adopt either a centralized production strategy, which has lower costs due to economies of scale, or a decentralized production strategy that is somewhat more expensive. However, the firm could also use transfer pricing to exploit the two markets' difference in selling seasons; as shown in Fig. 3.4, that strategy could offset the disadvantage of decentralized production (Kouvelis and Gutierrez 1997).

Allocation options refer to the distribution of products to specific markets in response to the realization of previously uncertain outcomes—for example, demand

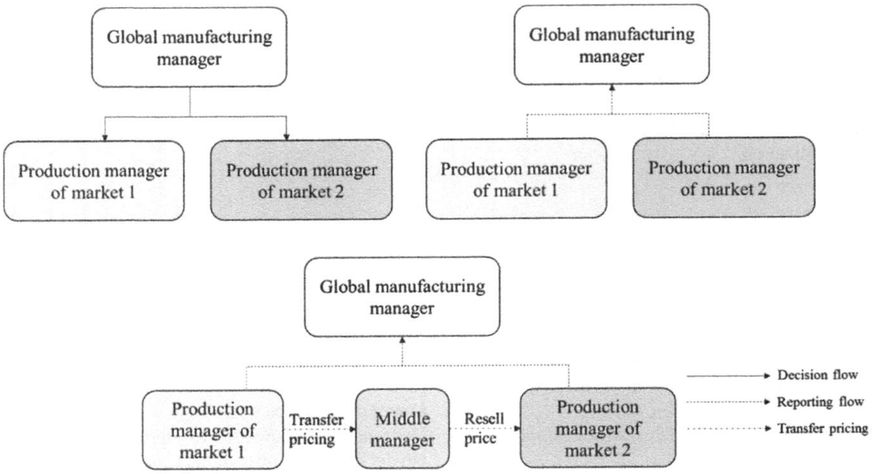

Fig. 3.4 Centralized production, decentralized production, and transfer pricing. Adapted from Kouvelis and Gutierrez (1997)

and the exchange rate (Ding et al. 2007). *After-sales services* are customer support and maintenance services that are offered subsequent to product purchase (Kim et al. 2007). *Entry/exit options* are the strategies of exploring new markets, holding back products, halting production, or exiting a market in order to mitigate demand risk and price risk.

(4) *Network flexibility* combines the three previous categories of operational hedging in a network structure. On the one hand, *supply chain network options* provide operational flexibility by altering production capacity and supply as well as processing and distribution linkages in a single-product setting; see Fig. 3.5. On the other hand, *network configuration* adjusts production networks that involve various supply, processing, and storage points in a multiple-product setting (Tomlin and Wang 2005; van Mieghem 2007).

Coordination options are mechanisms that align the incentives of supply chain members via contracting or information sharing. Various types of supply contracts are designed to coordinate a decentralized supply chain so that the performance benchmark of a centralized supply chain's profit can be more nearly achieved. One example is that of a wholesale price contract; here the unit price is fixed, the buyer retains the revenue, and any excess stock can be salvaged at a pre-specified price. A buy-back contract is a return policy under which the manufacturer "buys back", at a unit rate, an agreed-upon portion of the retailer's excess inventory. With an options contract, the buyer pays a reservation price up-front and executes the supply option by paying a per-unit execution fee. The only difference between an options contract and a buy-back contract with unlimited return is that the latter involves a two-way transportation process between supplier and buyer. Revenue-sharing

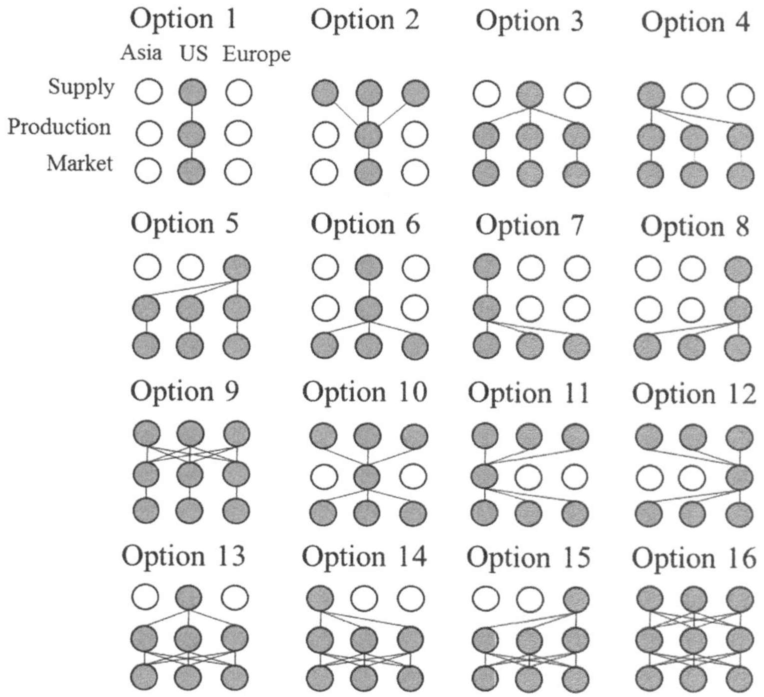

Fig. 3.5 Supply chain network options. Adapted from Huchzermeier and Cohen (1996)

contracts allow the retailer to benefit from a lower up-front wholesale price, although the retailer must remit to the manufacturer for each rental unit. Quantity-based contracts that feature a minimum order quantity enable buyers to adjust their orders (within a certain range) when necessary. Readers are referred to Cachon (2003) for an excellent review of supply chain contract analysis. Another coordination option is information sharing, which can minimize the "bullwhip effect" (i.e., demand variability increases from customers towards upstream supply) and thus optimize supply chain profit (Chen 2003). *Integration options* aim to synchronize information, material, and financial flows in a supply chain network through mergers and acquisitions (Nagurney 2009) or investment in information technology (IT) resources (Rai et al. 2006). The network flexibility enabled by *reverse logistics* involves the design, execution, and monitoring of a closed-loop supply chain over a product's life cycle—including the recovery of value from returns (Guide and Van Wassenhove 2009).

(5) *Timing flexibility* focuses on the time dimension of supply chain processes. Advance flexibility is defined as the strategic advancement of operational decisions for the purpose of matching product supply with demand. For instance, quick response enables fashion goods retailers to shorten lead times so

that a greater portion of production can be marketed in response to customer demand (Fisher and Raman 1996). Options contracts allow for the ex ante reservation and ex post execution of excess capacity (Donohue 2000). Besides, sourcing commodity products from both forward and spot markets gives firms the operational flexibility by differentiating the timing of quantity decisions (Kouvelis et al. 2013). Postponement flexibility amounts to the deferring of capacity investment, production quantity, and/or pricing decisions until after the realization of uncertainty (van Mieghem and Dada 1999). The firm can even alter a product's design to enable the postponement of differentiation—that is, the timing of when generic products are customized for specific markets (Lee 1996). For example, Hewlett-Packard (HP) redesigned its DeskJet printers by delaying the point of product differentiation. Hence HP now manufactures and ships generic printers to distribution centers in different regions, and those generic printers are then customized for the country-specific markets served by each center. Thus generic printers are produced (by HP) according to a make-to-stock system whereas the country-specific printers are customized in a make-to-order manner. This postponement strategy allows HP to respond quickly and effectively to changes in demand (Lee and Tang 1997).

(6) *Flexibility mix* is a mechanism for synchronizing the five previously discussed flexibilities. *Inter-category mix* combines flexibilities across different categories. In this way, four different supply chain structures can be identified depending on the extent to which postponement flexibility and modularization are adopted (Ernst and Kamrad 2000); see Fig. 3.6. *Intra-category mix* combines flexibilities within a given category. For instance, the strategies of multisourcing, contingent supply, and inventory mitigation can be integrated to mitigate the risk of supply disruptions (Tomlin 2006).

3.3 Decentralized Supply Risk Management

When a supply chain is centralized, the incentives of business units are well aligned to optimize value for the integrated firm. In practice, however, most supply chains consist of various profit-maximizing entities in a decentralized setting. Hence the misalignment of incentives among supply chain partners could lead to horizontal competition among suppliers and buyers as well as to vertical competition and information asymmetry between supplier and buyer. In the context of dual sourcing, supplier competition and diversification benefits may interact with each other in various ways. In order to illustrate these interaction effects, we next present a series of examples adapted from Aydin et al. (2012).

Example 3.1 Consider a supply chain consisting of one buyer and two potential suppliers. The buyer serves market demand of d units, and the product's unit price is r. The unit production cost is c_1 for supplier 1 and c_2 for supplier 2, where $c_1 < c_2$. The two suppliers are located in different markets and so do not engage in

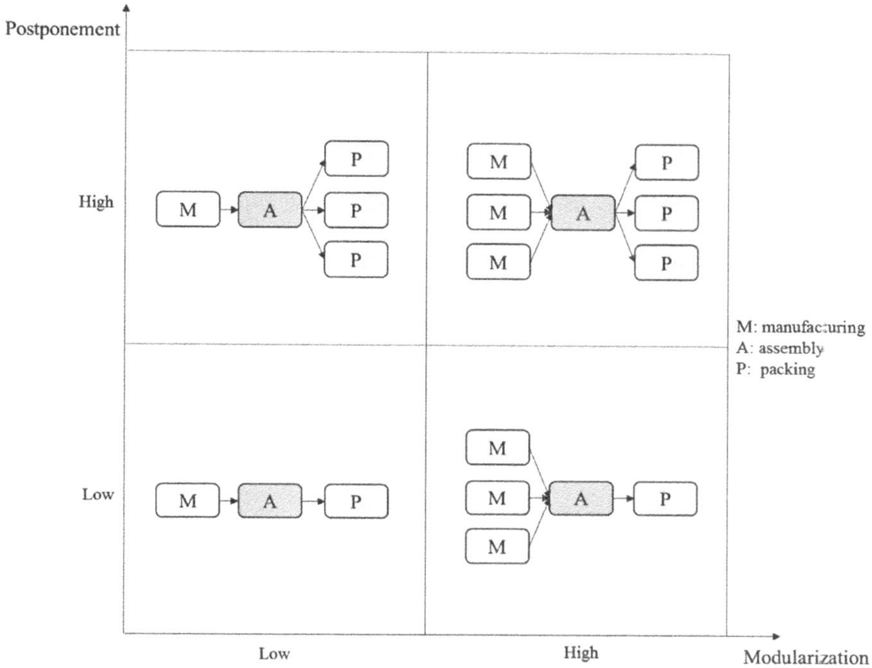

Fig. 3.6 Flexibility mix of modularization and postponement. Adapted from Ernst and Kamrad (2000)

price competition with each other. The probability of disruption is α for either supplier. In the event of disruption, the total order quantity will be unfulfilled; otherwise, the entire order quantity is served (this is known as the "all-or-nothing" case). The buyer has enough market power to set the wholesale prices w_1 and w_2 of (respectively) suppliers 1 and 2. Payment is required up-front. The reservation profit of each supplier is zero—that is, the suppliers will accept any wholesale price that does not exceed their respective production costs. In this example, it is optimal for the profit-maximizing buyer to set a wholesale price equal to the suppliers' respective production costs: $w_1 = c_1$ and $w_2 = c_2$.

Example 3.1a Here we assume a perfect and positive correlation between supplier disruptions; that is, the correlation ρ of supplier disruption is equal to 1. Thus the two suppliers' production lines are either both up or both down. It follows that the buyer does not benefit from placing orders with both suppliers—in other words, there is no value in diversification because disruptions are perfectly correlated. In this example, the buyer will choose the supplier that accepts a lower wholesale price; thus the buyer prefers to source from supplier 1, which has a lower production cost. The buyer's expected profit from placing an order quantity of q_1 with supplier 1 is

$$\pi_b^{1a} = (1-\alpha)r\min\{d,q_1\} + \alpha r\min\{d,0\} - c_1q_1.$$

So the optimal order quantities placed with these two suppliers are $q_1 = d$ and $q_2 = 0$, and the buyer's optimal expected profit can be written as $\pi_b^{1a*} = [(1-\alpha)r - c_1]d$.

Example 3.1b In contrast, now assume that supplier disruptions are entirely independent: the disruption correlation $\rho = 0$. Hence the likelihood of both suppliers' production lines being down is α^2, the likelihood that one supplier is down and the other up is $\alpha(1-\alpha)$, and the likelihood of both suppliers being up is $(1-\alpha)^2$. Then the buyer's expected profit when placing respective order quantities q_1 and q_2 to suppliers 1 and 2 is

$$\begin{aligned}\pi_b^{1b} = (1-\alpha)^2 r\,min\{d,q_1+q_2\} + \alpha(1-\alpha)r\min\{d,q_1\} \\ + \alpha(1-\alpha)r\min\{d,q_2\} + \alpha^2 r\min\{d,0\} - c_1q_1 - c_2q_2.\end{aligned}$$

In this case, if the buyer places order quantities of $q_1 = d$ and $q_2 = 0$ then its expected profit is $\pi_b^{1b} = [(1-\alpha)r - c_1]d$, or the same as in the optimal case of Example 3.1a. Since supplier disruptions are independent, it could be beneficial for the buyer to diversify. So if the buyer orders d units from each supplier $(q_1 = q_2 = d)$, then its expected profit is

$$\begin{aligned}\pi_b^{1b} &= (1-\alpha)^2 r\min\{d,2d\} + 2\alpha(1-\alpha)r\min\{d,d\} \\ &\qquad + \alpha^2 r \times \min\{d,0\} - c_1d - c_2d \\ &= \left[(1-\alpha)^2 r + 2\alpha(1-\alpha)r - c_1 - c_2\right]d = [(1-\alpha^2)r - c_1 - c_2]d.\end{aligned}$$

The value increment to the buyer's expected profit is $\Delta = \pi_b^{1b} - \pi_b^{1a*} = [(1-\alpha^2)r - c_1 - c_2]d - [(1-\alpha)r - c_1]d = \alpha(1-\alpha)r - c_2d$. Therefore, the buyer is better-off as long as the revenue increment exceeds the additional cost of dual sourcing; that is, if $\alpha(1-\alpha)r > c_2d$. This value increment captures the *diversification benefit* of dual sourcing relative to the optimal profit when sourcing only from supplier 1.

In sum, if supplier disruptions are *not* perfectly correlated then the buyer should adopt multisourcing (here, dual sourcing) instead of single sourcing in order to benefit from diversification. We remark that risk management and diversification can be valuable irrespective of assumptions made with regard to risk aversion. When the objective function of maximization (resp. minimization) is concave (resp. convex), the benefit of diversification will be evident (Froot et al. 1993). The value of diversification decreases with an increase in ρ, the correlation of supplier disruptions. In analogy to the diversification benefit in financial portfolio management, there is no benefit from operational diversification when supplier disruptions are perfectly positively correlated.

Example 3.2 We now assume, ceteris paribus, that the suppliers operate in one market and therefore compete with each other when setting prices.

Example 3.2a If supplier disruptions are perfectly positively correlated (i.e., if $\rho = 1$), then both suppliers' production lines are either up or down simultaneously. Hence, because there is then zero diversification benefit, the buyer chooses single sourcing from the supplier bidding the lower wholesale price. In Bertrand competition, suppliers compete for the buyer's total order quantity by bidding lower wholesale prices. In this setting, supplier 1 will win the bid at a wholesale price just below c_2—that is, at $w_1 = \lim_{\Delta \to 0}(c_2 - \Delta)$ for $\Delta > 0$. Supplier 2 cannot match this wholesale price because it is not profitable (owing to 2's higher unit production cost c). The buyer's expected profit from ordering quantity q_1 from supplier 1 is

$$\pi_b^{2a} = (1-\alpha) r \min\{d, q_1\} + \alpha r \min\{d, 0\} - w_1 q_1.$$

The optimal order quantities to each supplier are thus $q_1 = d$ and $q_2 = 0$, and the buyer's optimal expected profit $\pi_b^{2a*} = \lim_{\Delta \to 0}[(1-\alpha)r - c_2 + \Delta]d$. In comparison with Example 3.1a, the buyer's profit decreases by $\pi_b^{1a*} - \pi_b^{2a*} = \lim_{\Delta \to 0}(c_2 - c_1 - \Delta) > 0$. This decline in the buyer's expected profit is due to the Bertrand competition, whereby supplier 1 has an inflated wholesale price.

Example 3.2b If supplier disruptions are uncorrelated (i.e., if $\rho = 0$), then the buyer's optimal order quantities depend on the suppliers' wholesale prices. Here the buyer is indifferent between dual sourcing and single sourcing when both suppliers bid wholesale prices at $w_1 = w_2 = r\alpha(1-\alpha)$. Hence the equilibrium wholesale price in this Bertrand competition is $w_1 = w_2 = r\alpha(1-\alpha)$, which indicates the value of a supplier that is up while its rival is down (see Fig. 3.7). The intuition underlying this equilibrium wholesale price is as follows. On the one hand, the buyer will not source from any supplier that bids a price higher than $r\alpha(1-\alpha)$, or $r\alpha(1-\alpha) + \Delta$ for $\Delta > 0$. On the other hand, if one supplier bids a price lower than $r\alpha(1-\alpha)$, or $r\alpha(1-\alpha) - \Delta$ for $\Delta > 0$, then the supplier will receive the same order quantity but will earn Δ less in expected profit. Thus neither supplier has an incentive to deviate from the equilibrium wholesale price.

Given this equilibrium wholesale price, supplier 1's expected profit is $\pi_{s1}^{2b} = [r\alpha(1-\alpha) - c_1]q_1$, supplier 2's expected profit is $\pi_{s2}^{2b} = [r\alpha(1-\alpha) - c_2]q_2$, and the buyer's expected profit with respective order quantities q_1 and q_2 to suppliers 1 and 2 is

$$\begin{aligned}\pi_b^{2b} &= (1-\alpha)^2 r \min\{d, q_1 + q_2\} + \alpha(1-\alpha) r \min\{d, q_1\} \\ &\quad + \alpha(1-\alpha) r \min\{d, q_2\} + \alpha^2 r \min\{d, 0\} - r\alpha(1-\alpha)(q_1 + q_2).\end{aligned}$$

Fig. 3.7 Equilibrium wholesale prices in Bertrand competition

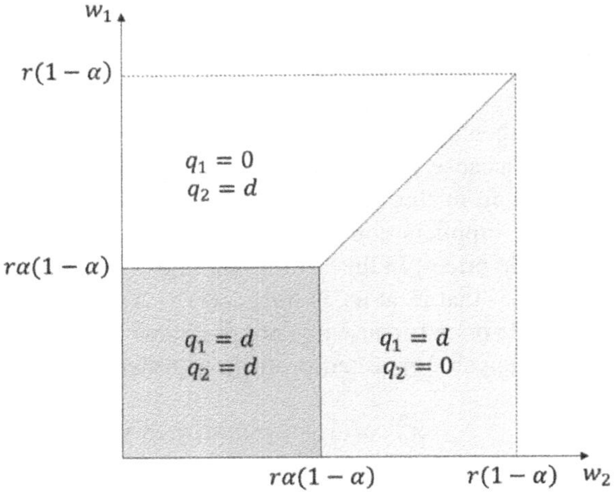

Table 3.2 Interaction between supplier competition and disruption correlation

Example	Correlation	Competition	Expected profit
3.1a	✓		1200
3.1b			1290
3.2a	✓	✓	1170
3.2b		✓	1080

In order to diversify its supply base, the buyer orders d units from each supplier ($q_1 = q_2 = d$); then its expected profit is $\pi_b^{2b} = (1-\alpha)^2 rd$. The value change in the buyer's expected profit is $\Delta = \pi_b^{2b} - \pi_b^{2a*}$ $\Delta = \pi_b^{2b} - \pi_b^{2a*} = (1-\alpha)^2 rd - \lim_{\Delta \to 0}[(1-\alpha)r - c_2 + \Delta]d$, which establishes that the value of diversification depends on how the buyer's expected profits when dual sourcing compare with its expected profits when sourcing only from supplier 1.

To illustrate these examples numerically, we set the values of parameters in Examples 3.1 and 3.2 to $c_1 = 60$, $c_2 = 63$, $p = 300$, $\alpha = 0.4$, and $d = 10$. Then the buyer's expected profits are $\pi_b^{1a*} = 1200$, $\pi_b^{1b} = 1290$, $\pi_b^{2a*} = 1170$, and $\pi_b^{2b} = 1080$.

These outcomes are summarized in Table 3.2. From Example 3.1a to 3.1b, when the correlation of supplier disruption decreases from 1 to 0, the benefit of supplier diversification increases by 90 (i.e., in the absence of supplier price competition). However, when we move from Example 3.2a to 3.2b, the diversification benefit is (more than) offset by supplier price competition and so the buyer's expected profit decreases by 90. This result shows that supplier diversification and price competition can have the opposite effects. In Example 3.2, the presence of price competition increases the supplier's wholesale price from the production cost to $r\alpha(1-\alpha)$—a price inflation effect that dominates the diversification benefit. In short: the value of diversification observed in traditional financial portfolio optimization applies in Example 3.1 but is reversed, because of supplier price competition, in Example 3.2.

Example 3.3 Consider a supply chain consisting of one buyer and two suppliers. The two suppliers compete in a total-cost, reverse-English auction to win the buyer's order. The suppliers' unit production costs, c_1 and c_2, are not known to the buyer, but it does know the respective unit transportation costs t_1 and t_2. In the total-cost reverse-English auction, the suppliers bid their wholesale prices w_1 and w_2 in turn; then the buyer compares the total unit costs (i.e., after accounting for transportation costs). The total costs are visible to both suppliers. If no supplier is willing to bid a lower wholesale price, then the bidder of the lower total cost wins the buyer's ordering contract.

Example 3.3a We now assume that the buyer is based in Germany and that both suppliers are from China; in this case, the unit transportation costs for both suppliers are the same ($t_1 = t_2$). As in our previous examples, the unit production cost of supplier 1 is lower than that of supplier 2: $c_1 < c_2$. Therefore, supplier 1 will win the bid at a wholesale price just below the production cost of supplier 2; that is, at $w_1 = \lim_{\Delta \to 0}(c_2 - \Delta)$ for $\Delta > 0$. The total cost for the buyer is then $TC_b^{3a} = w_1 + t_1 = \lim_{\Delta \to 0}(c_2 - \Delta) + t_2$.

Example 3.3b In order to diversify its supplier base and reduce transportation costs, the German buyer now runs the total-cost reverse-English auction with suppliers located in different countries: supplier 1 is from Eastern Europe and supplier 2 is from China. In this case, the transportation cost for supplier 1 is lower than that for supplier 2 ($t_1 < t_2$). Since supplier 1's production cost is lower than supplier 2's ($c_1 < c_2$), it follows that the total cost of sourcing from supplier 1 is lower than that from supplier 2; that is, $TC_1 = w_1 + t_1 < TC_2 = w_2 + t_2$. In the total-cost reverse-English auction, supplier 1 will win by bidding a wholesale price that leads to a total cost just below the lowest total cost that supplier 2 could offer—namely, $w_1 = \lim_{\Delta \to 0}(c_2 + t_2 - \Delta) - t_1$ for $\Delta > 0$. Hence the buyer's total unit cost is $TC_b^{3b} = w_1 + t_1 = \lim_{\Delta \to 0}(c_2 + t_2 - \Delta)$.

In comparison with Example 3.3a, the buyer's total cost remains the same in spite of the diversified supplier base. Although the transportation cost is reduced by replacing a Chinese supplier with an Eastern Europe supplier, the resulting cost advantage is offset by the price competition in the total-cost reverse-English auction, in which supplier 1 can inflate its wholesale price *provided that* the unit total cost remains lower than that of supplier 2. Similarly to Example 3.2, the firm's benefit from diversifying its suppliers is offset by the price competition between them. So whenever one supplier has a higher transportation cost, the other can raise the wholesale price to enjoy a "windfall" benefit—to the buyer's detriment.

Examples 3.1–3.3 establish that supplier diversification and price competition can work against each other in the buyer's profit optimization. This outcome raises an interesting question: Is there some way for a buyer to reap not only the benefit of supplier diversification but also a price advantage due to supplier competition? The answer is Yes—at least for buyers that establish a *four*-supplier base that consists of two suppliers each in two geographical clusters. Within a given cluster, supplier disruptions will exhibit a strong correlation: suppliers from one region or country are likely to experience the same natural hazards, financial risks, and political instability; at the same time, these suppliers compete with each other on wholesale price. Yet because correlation of supplier disruptions in different clusters is *not* strong, the buyer gains the price advantage due to Bertrand competition.

References

Aydın, G., Babich, V., Beil, D. R., & Yang, Z. (2012). Decentralized supply risk management. In: Kouvelis, P., Boyabatli, O., Dong, L., & Li, R. (Eds.), *Handbook of integrated risk management in global supply chains* (389–424). New York: Wiley.

Babich, V., Burnetas, A. N., & Ritchken, P. H. (2007). Competition and diversification effects in supply chains with supplier default risk. *Manufacturing & Service Operations Management, 9,* 123–146.

Blome, C., & Henkel, M. (2009). Single versus multiple sourcing: A supply risk management perspective. In Zsidisin, G. A., & Ritchie, B. (Eds.), *Supply chain risk a handbook of assessment, management, and performance*. Berlin: Springer.

Cachon, G. (2003). Supply chain coordination with contracts. In Graves, S., de Kok, T. (Eds.), *Handbooks in operations research and management science: Supply chain management* (pp. 229–340). North Holland, Amsterdam.

Cachon, G. (2003). Supply chain coordination with contracts. In Graves, S., de Kok, T. (Eds.), *Handbooks in operations research and management science: Supply chain management* (pp. 341–422). North Holland, Amsterdam.

Chod, J., & Rudi, N. (2005). Resource flexibility with responsive pricing. *Operations Research, 53* (3), 532–548.

Christopher, M. (1992). *Logistics and supply chain management*. London: Pitman.

Dada, M., Petruzzi, N. C., & Schwarz, L. B. (2007). A newsvendor's procurement problem when suppliers are unreliable. *Manufacturing & Service Operations Management, 9,* 9–32.

Ding, Q., Dong, L., & Kouvelis, P. (2007). On the integration of production and financial hedging decisions in global markets. *Operations Research, 55,* 470–489.

Donohue, K. (2000). Efficient supply contracts for fashion goods with forecast updating and two production modes. *Management Science, 46,* 1397–1411.

Ernst, R., & Kamrad, B. (2000). Evaluation of supply chain structures through modularization and postponement. *European Journal of Operational Research, 124,* 495–510.

Fisher, M., & Raman, A. (1996). Reducing the cost of demand uncertainty through accurate response. *Operations Research, 44,* 87–99.

Froot, K. A., Scharfstein, D. S., & Stein, J. C. (1993). Risk management: Coordinating corporate investment and financing policies. *Journal of Finance, 48,* 1629–1658.

Gamba, A., & Triantis, A. J. (2014). Corporate risk management: Integrating liquidity, hedging, and operating policies. *Management Science, 60,* 246–264.

Guide, V. D. R., Jr., & Van Wassenhove, L. N. (2009). The evolution of closed-loop supply chain research. *Operations Research, 57,* 10–18.

Huchzermeier, A., & Cohen, M. A. (1996). Valuing operational flexibility under exchange rate risk. *Operations Research, 44,* 100–113.

Kazaz, B., Dada, M., & Moskowitz, H. (2005). Global production planning under exchange-rate uncertainty. *Management Science, 51,* 1101–1119.

Kim, S.-H., Cohen, M., & Netessine, S. (2007). Performance contracting in after-sales service supply chains. *Management Science, 53,* 1843–1858.

Kouvelis, P. (1999). Global sourcing strategies under exchange rate uncertainty. In Tayur, S., Ganeshan, R., Magazine, M. (Eds.), *Quantitative models for supply chain management* (625–668). Dordrecht: Kluwer Academic Publishers.

Kouvelis, P., & Gutierrez, G. (1997). The newsvendor problem in a global market: Optimal centralized and decentralized control policies for a two-market stochastic inventory system. *Management Science, 43,* 571–585.

Kouvelis, P., Li, R., & Qing, D. (2013). Managing storable commodity risks: The role of inventories and financial hedge. *Manufacturing & Service Operations Management, 15,* 507–521.

Latour, A. (2001). Trial by fire: A blaze in Albuquerque sets off major crisis for cell-phone giants. *Wall Street J.,* January 29, 2001.

Lee, H. L. (1996). Effective management of inventory and service through product and process redesign. *Operations Research, 44,* 151–159.

Lee, H. L., & Tang, C. S. (1997). Modeling the costs and benefits of delayed product differentiation. *Management Science, 43,* 40–53.

Nagurney, A. (2009). A system-optimization perspective for supply chain network integration: The horizontal merger case. *Transportation Research, 45,* 1–15.

Pindyck, R. S., & Rubinfeld, D. L. (1989). *Microeconomics*. New York: Macmillan.

Rai, A., Patnayakuni, R., & Seth, N. (2006). Firm performance impacts of digitally enabled supply chain integration capabilities. *MIS Quarterly, 30,* 225–246.

Roszkowski, M. J., & Davey, G. (2010). Risk perception and risk tolerance changes attributable to the 2008 economic crisis: A subtle but critical difference. *Journal of Financial Service Professionals, 64,* 42–53.

Sting, F. J., & Huchzermeier, A. (2010). Ensuring responsive capacity: How to contract with backup suppliers. *European Journal of Operational Research, 207,* 725–735.

Tang, C. S. (2006). Perspectives in supply chain risk management. *Internationl Journal of Production Economics, 103,* 451–488.

Tomlin, B. (2006). On the value of mitigation and contingency strategies for managing supply chain disruption risks. *Management Science, 52,* 639–657.

Tomlin, B., & Wang, Y. (2005). On the value of mix flexibility and dual sourcing in unreliable newsvendor networks. *Manufacturing & Service Operations Management, 7,* 37–57.

van Mieghem, J. A. (2007). Risk mitigation in newsvendor networks: Resource diversification, flexibility, sharing, and hedging. *Management Science, 53,* 1269–1288.

van Mieghem, J. A. (2012). Risk management and operational hedging: An overview. In P. Kouvelis, L. Dong, O. Boyabatli, & R. Li (Eds.), *Handbook of integrated risk management in global supply chains* (pp. 13–49). New York: Wiley.

van Mieghem, J. A., & Dada, M. (1999). Price versus production postponement: Capacity and competition. *Management Science, 45,* 1631–1649.

Vickery, S., Calantone, R., & Dröge, C. (1999). Supply chain flexibility: An empirical study. *Journal of Supply Chain Management, 35,* 16–24.

Wang, Y., Gilland, W., & Tomlin, B. (2010). Mitigating supply risk: Dual sourcing or process improvement? *Manufacturing & Service Operations Management, 12,* 489–510.

Yang, Z., Aydin, G., Babich, V., & Beil, D. (2009). Supply disruptions, asymmetric information, and a backup production option. *Management Science, 55,* 192–209.

Zsidisin, G. A. (2005). Approaches for managing price volatility. In *5th International Research Seminar on Risk and the Supply Chain*. Centre for Logistics and Supply Chain Management, School of Management, Cranfield University, 12th September 2005.

Chapter 4
Integrated Risk Management in Multinational Corporations

This chapter synthesizes the conceptual framework and empirical investigation of IRM (integrated risk management). We start by comparing frameworks of IRM and ERM (enterprise risk management). Next, we discuss risk attitudes and objective formulations in quantitative optimization. Finally, we present various risk measures and classify risk management strategies according to their treatment of risk.

4.1 Integrated Risk Management: Concepts and Frameworks

The closed-loop view of operations–finance interfaces (see Sect. 1.1) results in two concepts of risk management: the inter-organizational *supply chain risk management*, discussed in Chap. 3; and the cross-functional *enterprise risk management*.

In this section, we elaborate our framework for integrated risk management and compare the conceptual framework of IRM with that of ERM. Recall that we defined *integrated risk management* as "the joint analysis, synthesis, and optimization of operational and financial risk management across functional units in an enterprise and across supply chain partners." In contrast, *enterprise risk management* is defined as "a process, effected by an entity's board of directors, management and other personnel, applied in strategy setting and across the enterprise, designed to identify potential events that may affect the entity, and manage risk to be within its risk appetite, to provide reasonable assurance regarding the achievement of entity objectives" (COSO 2004). A more concise definition is offered by Chapman (2006), who describes ERM as "a systematic process, embedded in a company's system of internal control (spanning all business activity), to satisfy policies effected by its board of directors, aimed at fulfilling its business objectives and safeguarding both the shareholder's investment and the company's assets."

L. Zhao and A. Huchzermeier, *Supply Chain Finance*, EURO Advanced Tutorials on Operational Research, https://doi.org/10.1007/978-3-319-76663-8_4

Integrated risk management starts with *input*, which includes risk management objectives, risk data and information about the enterprise and the supply chain environment. There are two types of objectives in risk management: *strategic objectives*, which are long-term goals (set by C-level management) that accord with corporate vision and the creation of shareholder value; and *tactical objectives*, which are short-term goals (based on performance measures) that ensure the firm's operations and finance are on track. Tactical objectives are usually specified in terms of such key performance indicators (KPIs) as inventory turnover, service level, sales revenue, return on investment, and so forth. Risk data are collected to assess exposure to risk as well as its likelihood and potential impact; such data provide the basis for quantitative risk management. The firm's environment is described in terms of supply chain linkages, the competitive landscape, the company's market share, its organizational structure, and also industry standards and regulations. All this information is collected and updated on a continuous basis so that the firm can detect and respond in a timely manner to changes in the sources of uncertainty, in its environment, and in its objectives.

Risk identification involves the scanning, classification, and recording of sources of uncertainty. These sources of uncertainty can be grouped into three levels: contextual uncertainty, supply chain uncertainty, and firm-specific uncertainty (cf. Miller 1992). *Contextual* uncertainty is driven by several factors: political instability or turmoil; uncertain government policy; macroeconomic variables, such as inflation, interest rates, and exchange rates; social factors, including terrorism and protest; and natural disasters. *Supply chain* uncertainty reflects factor market uncertainty, product market uncertainty, and competition factors associated with current rivals and new entrants. *Firm-specific* uncertainty refers to risks associated with internal operations, environmental risks associated with the firm's own products, innovation risk in the R&D process, credit risk, and uncertainties about stakeholder behavior. All these sources of uncertainty affect the firm and thereby trigger operational and financial risks (see Table 1.1). After those operational and financial risks are identified, three types of interdependence can be detected among them (see Sect. 1.3). The firm can then examine ten dimensions of integration—in categories of feasibility, trade-offs, and structure—to determine when it is optimal to synchronize operations and finance (Sect. 1.4). That examination enables it to select operational hedging and financial flexibility strategies as a function of the risks that have been identified and the integration conditions that have been satisfied (Sect. 1.5). Finally, the integrated optimization of operations and finance involves conducting a relationship analysis (Sect. 1.6) and approach choice (Sect. 2.1).

Figure 4.1 presents a four-step "waterfall" model of integrated risk management, a model whose components are interconnected by an iterative process of verification and modification. This process ensures that the realization of risk management processes are consistent with the original design and enables adjustment of the initial plan in response to immediate feedback from the next step. Once the risk management strategies are implemented through deployment across functional units, the entire process is monitored and reviewed (in terms of selected performance metrics) at various organizational levels. Continuous improvement is

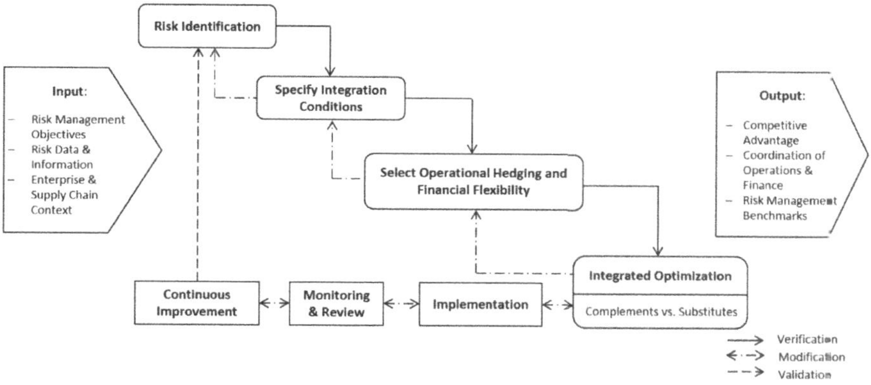

Fig. 4.1 Framework for integrated risk management

connected to risk identification by validation, or the updating of input parameters on a regular basis. The framework of integrated risk management illustrated in Fig. 4.1 is meant to generate outcomes that include well-coordinated operations–finance interactions leading to competitive advantages for the firm and the successful achievement of risk management goals.

Figure 4.2 illustrates the enterprise risk management framework devised by the Committee of Sponsoring Organizations of the Treadway Commission (COSO). This integrated framework is visualized as a three-dimensional cube generated by four categories of *objectives* (strategic, operations, reporting, and compliance), the vertical columns; eight components of the risk *management process*, the horizontal rows; and four *organizational levels* (from entity through subsidiary), the third dimension (front to back). This framework focuses on the entirety, consistency, and strategy "cascading" of a firm's enterprise risk management. Here the risk management process includes the internal environment, objective setting, event identification, risk assessment, risk response, control activities, information and communication, and monitoring. Each organizational unit aims to enhance value even as it faces uncertainty. The goal of ERM is to optimize when and how much uncertainty to accept, share, or mitigate toward the end of increasing stakeholder value.

In Fig. 4.3 we show an alternative ERM framework proposed in Chapman (2006). This framework views enterprise risk management as an iterative process for risk and opportunity management that comprises four primary functions: (i) policy formulation to create vision, mission, value, and culture; (ii) strategic thinking to set the firm's direction by evaluating its internal resources and external market conditions; (iii) supervisory management to monitor and review the implementation process; and (iv) accountability to stakeholders.

When comparing our integrated risk management framework with the general risk management framework of ISO and the ERM frameworks of COSO and Chapman (2006), we see that our integrated framework emphasizes evaluating the

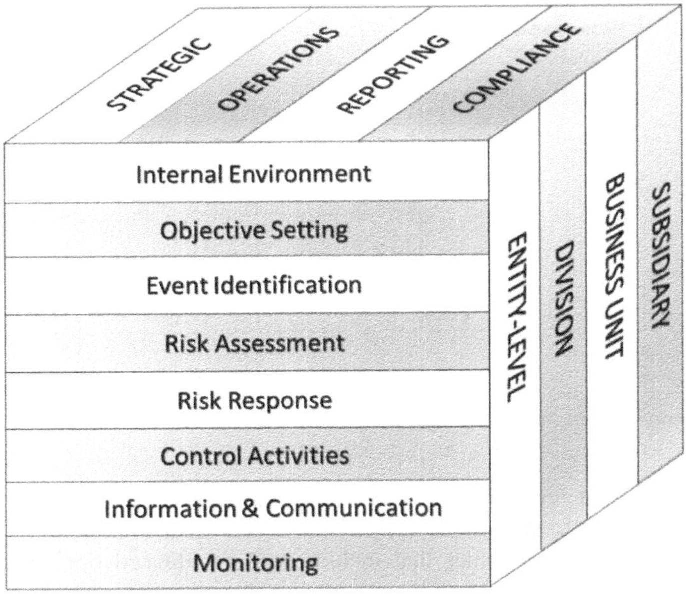

Fig. 4.2 ERM framework of COSO (Adapted from COSO 2004)

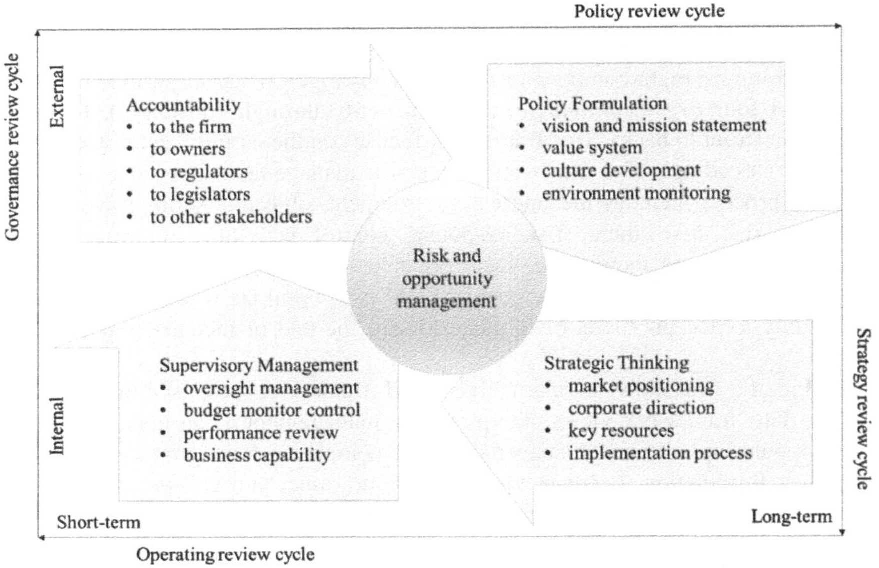

Fig. 4.3 Alternative ERM framework (Adapted from Chapman 2006)

interdependence of operations and finance so that the two functions can be *jointly* optimized when the integration conditions are fulfilled. In contrast, the enterprise risk management frameworks focus on policy formulation and strategy cascading at the corporate, business unit, functional, and subsidiary levels.

4.2 Integrated Risk Management: Objectives and Measures

Risk is associated with sources of uncertainty and the notion of randomness, aspects that are central to probability theory. (We assume that the readers of this book are familiar with the basic notions of probability and statistics.) Various risk measures and utility functions have been adopted in quantitative integrated risk management. Two frequently used measures are the variance and the standard deviation:

$$\text{Variance} = \mathbb{E}[\bar{X} - X]^2 = \delta^2;$$
$$\text{Std.Dev.} = \delta = \sqrt{\mathbb{E}[\bar{X} - X]^2}.$$

Both of these risk measures are symmetric in that they treat upside potential and downside risk equally. In contrast, downside risk measures capture only the undesirable consequences of uncertainty. Three such measures (see Nawrocki 1999) are:

$$\text{below-mean semivariance} = \mathbb{E}[(\bar{X} - X)^+]^2,$$
$$\text{below-target } t \text{ semivariance} = \mathbb{E}[(t - X)^+]^2,$$
$$\text{expected below-target } t \text{ risk} = \mathbb{E}[t - X]^+;$$

here $X^+ = \max\{X, 0\}$. Another popular downside risk measure is value-at-risk (VaR), which defines the maximum loss that the focal firm could incur at a given confidence level (e.g., 95% or 99%) over a selected period.

We use $(\Omega; \mathbb{F}; P)$ to denote a probability space Ω with filtration $\mathbb{F}$ and probability measure P. A loss function L based on uncertain factors is written as $L : \Omega \to \mathbb{R}$. Suppose that the random vector $\boldsymbol{y}$ is governed by P and is independent of the decision variable vector $\boldsymbol{x}$. For a given vector $\boldsymbol{x}$, the probability that a loss function $l(\boldsymbol{x})$ does *not* exceed the threshold α is $\Psi(\boldsymbol{x}, \alpha) \triangleq \int_{l(\boldsymbol{x}) \leq \alpha} f(\boldsymbol{y}) d\boldsymbol{y}$. For a given confidence level $\beta \in (0, 1)$ and a fixed decision vector $\boldsymbol{x}$, the value-at-risk is defined as

$$\text{VaR}_\beta(x, \boldsymbol{y}) \triangleq \min\{\alpha \in \mathbb{R} \sim \Psi(\boldsymbol{x}, \alpha) \geq \beta\}.$$

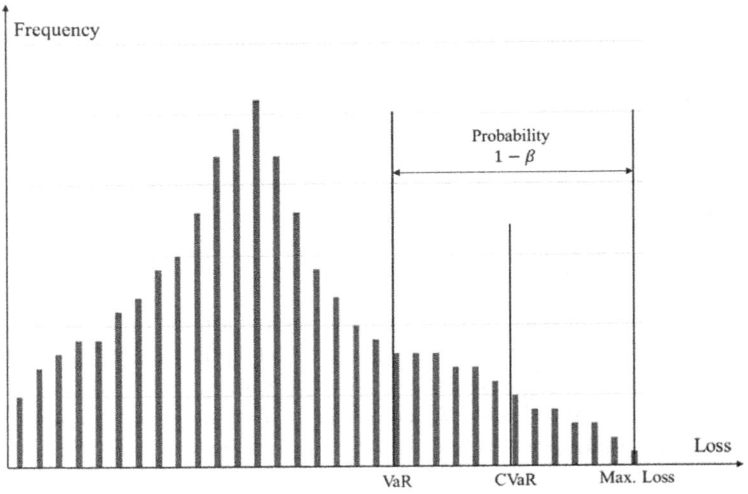

Fig. 4.4 Relationship between VaR and CVaR

Suppose, for instance, that the VaR for exchange rate risk is defined by the following parameters: 95% confidence level, 3% expected loss, and one-year time period. In this example, if a firm invests 100 million in the foreign exchange market then there is a 95% chance that the investment will not, over the next year, lose the firm more than 3 million. In other words, the VaR of this portfolio at 5% (i.e., 100% minus 95%) is 3 million for the period specified.

A closely related downside risk measure is conditional value-at-risk (CVaR), which can be viewed as a remedy of VaR. Conditional value-at-risk is the *average* of worst-case losses below the VaR threshold under certain conditions. If the confidence level is held constant, then VaR is the lower bound for CVaR. Formally, we have

$$\mathrm{CVaR}_\beta(\boldsymbol{x},\boldsymbol{y}) \triangleq \frac{1}{1-\beta}\int_{l(\boldsymbol{x},\boldsymbol{y})\geq \mathrm{VaR}_\beta(\boldsymbol{x},\boldsymbol{y})} l(\boldsymbol{x},\boldsymbol{y})f(\boldsymbol{y})d\boldsymbol{y}.$$

Figure 4.4 shows the relationship between VaR and CVaR along with their respective deviations from the expected value of a distribution.

The risk measures described here can be used to conduct a risk–return analysis based on various formulations of the focal firm's objectives. For this purpose, we combine a return function with a risk measure weighted by a coefficient of risk aversion. Thus, for example, a mean variance (MV) objective function is defined as

$$\mathrm{MV} = \mu - \frac{\gamma}{2}\sigma^2.$$

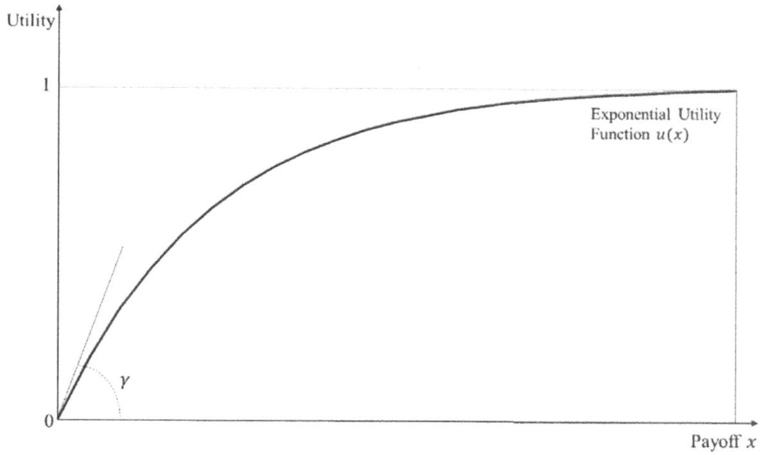

Fig. 4.5 Concavity of the exponential utility

Here μ is the expected value, and $\gamma > 0$ is the coefficient of absolute risk aversion. As γ increases, the MV objective becomes more sensitive to variance. Mean variance preferences are at the core of modern financial portfolio management and have inspired operations strategies for risk mitigation; the idea was first formulated in 1952 by Nobel laureate Harry Markowitz. The MV objective has two key benefits. First, it can be readily implemented based on measurable parameters. Second, it yields "good recommendations" even when there is ambiguity in the decision maker's utility function (van Mieghem 2003).

When the payoff function is normally distributed, the mean variance objective is equivalent to the (more accurate formulation of) expected, exponential utility:

$$u(x) = 1 - e^{-\gamma x}.$$

As the coefficient γ of absolute risk aversion increases, the exponential utility becomes more concave and thus more sensitive to downside deviations. This dynamic is plotted in Fig. 4.5. We can see that, with increasing concavity of the exponential utility, a risk-*averse* decision maker more strongly prefers a unit reduction in downside risk to a unit increase in upside potential. In contrast, a risk-*neutral* decision maker has a linear utility function and so is concerned only with the expected value of actual outcomes, such as the value of profit and cost functions.

4.3 Integrated Risk Management: Strategy Formulation in Multinational Corporations

In this section, we present a strategy formulation framework for integrating operations and finance in multinational corporations (MNCs) by linking the theories of organization economics and strategic management. Table 4.1 shows our strategic framework for integrated operations–finance risk management, along three dimensions, in terms of five processes and their corresponding approaches.

The three dimensions—incentives, framing, and mechanism—explore three fundamental questions: (1) Why should MNCs adopt integrated risk management? (2) What operational strategies and financial instruments should be employed based on the MNC's resources? (3) How can the integrated optimization of operations and finance be implemented?

(1) An MNC's integrated risk management is *incentivized* by the coexistence and interdependence of operational and financial risks, as detailed in Sect. 1.3. The macro-environmental factors can be identified in a PEST (political, economic, socio-cultural and technological) analysis (Schmieder-Ramirez and Mallette 2015), while Porter's five forces model can be adopted to analyze micro-environmental factors in industrial competition (Porter 1985).

(2) Integrated risk management in the MNC context can be *framed* in terms of the firm's resources. According to the resource-based view, the goal should be to build sustainable competitive advantages by evaluating whether the focal firm's key resources satisfy the criteria of value, rareness, inimitability, and non-substitutability. In other words, the firm's resources should dictate a value-creating strategy, be rare (if not unique), and be such that they can neither be duplicated nor substituted by competitors (Wernerfelt 1984; Barney 1991). Recall from Fig. 1.1 that a closed-loop view connecting material, monetary, and information flows underlies a useful framing strategy for optimizing operations–finance integration. Given the firm's choice of operational strategies and financial instruments, a relationship matrix can be used to assess whether its resources are complements or substitutes.

(3) The *mechanism* of risk management acts on organizational structure in that the firm's activities can be centralized or decentralized based on its choice of strategy. Figure 4.6 classifies the MNC environment into four types depending on the extent of global integration and local responsiveness: a *transnational* environment, where both factors are high; a *global* environment, where only global integration is high; a *multinational* environment, where only local responsiveness is high; and an *international* environment, where both factors are low. Another aspect of risk management mechanisms is the corporate–division relationship. Such relations can be categorized by governance

Table 4.1 Strategy formulation for integrating an MNC's operations and finance

Dimension	Incentives (Why)		Framing (What)		Mechanism (How)
Process	Environment	Industry	Resource	Strategy	Structure and activity
Approach	PEST analysis, General environmental uncertainties	Porter's five forces, Industry uncertainties	Resource-based view, Closed-loop view	Integration of operations & finance, Relationship matrix	Centralization/decentralization, Global integration/local responsiveness
References	Schmieder-Ramirez and Mallette (2015), Miller (1992)	Porter (1985), Miller (1992)	Barney (1991), Chap. 1	Tables 2.3 and 3.1	Table 4.2, Fig. 4.7, Ghoshal and Nohria (1993)

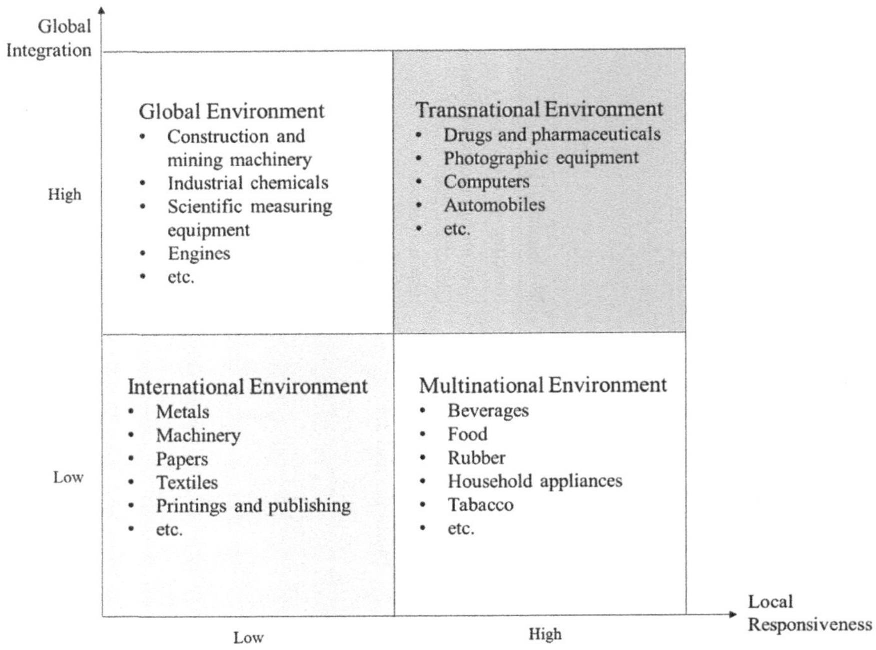

Fig. 4.6 Classification of MNC environments (Adapted from Ghoshal and Nohria 1993)

schemes: (i) centralization by established authority and hierarchy, (ii) formalization via bureaucratic regulations and procedures, or (iii) normative integration through the socialization of managers with shared goals and aligned incentives (Ghoshal and Nohria 1993).

Figure 4.7 illustrates a multinational corporation's functional structure for the integrated risk management of operations and finance. The accounting and finance departments should manage capital structure, accounting, and treasury by centralization with divisional audit. Asset liability management is directly connected to the creation of shareholder value in the long term by sustainable competitive advantages based on the firm's key resources. The treasury department manages financial hedging, via currency derivatives, based in part on shared information about the firm's operating cash flows. The accounting department controls, audits, and reviews the firm's operational performance using financial metrics; thus it is connected to the operations and supply chain division by real investment in capital budgeting and by revenue management in marketing and sales processes. The cycling process of material, monetary, and information flows between operations and finance is represented by the counterclockwise arrow signifying a closed loop (cf. Fig. 1.1). The operations and supply chain division manages capacity investment, production planning, and distribution logistics by way of decentralization and coordination. In the upstream supply chain, the management of supplier

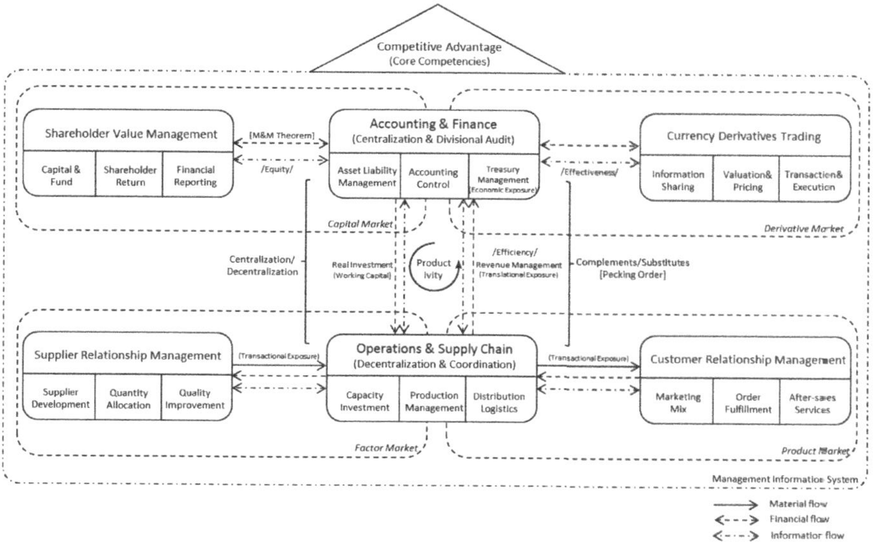

Fig. 4.7 The MNC's functional structure and stakeholder relationships in integrated risk management

relationships is connected to quantity decisions in capacity and production planning; downstream, the management of customer relationships connects product distribution with marketing, sales, and services.

The MNC is exposed to various sources of uncertainty stemming from operational activities and financial variables. In corporate risk management there are three categories of risk, of which the first is *credit* risk—that is, the risk of not receiving outstanding payments because the borrower defaults. The second risk category is *market* risk, or the variance in portfolio value due to changes in such market factors as stock and bond prices, interest and exchange rates, commodity prices, et cetera. In currency risk management, there are three types of exchange rate exposure. (i) Economic exposure is the extent to which a firm's present value of future operating cash flows is affected by exchange rate fluctuations; such exposure can be mitigated by financial hedging via currency derivatives in treasury management. (ii) Transaction exposure is the effect of exchange rate movements on contractual cash flows associated with receivables (exports) and payables (imports). This type of exchange rate exposure can be managed not only by financial hedging but also by currency invoicing, the leading and lagging of receipts and payments, and/or the firm's exposure netting. (iii) Translation exposure is defined as the effect of exchange rates on financial reporting when consolidating the assets and liabilities of foreign subsidiaries into the parent company's balance sheet.

Corporate risk management's third risk category is *operational* risk, which is the risk of losses resulting from internal operational activities, staff and systems, and/or external events. In operational risk management, the "three Es" of efficiency, effectiveness, and equity can be used as performance criteria. "Efficiency" in this context is the ratio of output to input; thus a greater volume of output for a given volume of input corresponds to greater efficiency. "Effectiveness" is a term describing the extent to which the firm's objectives are achieved (Otley 1999). So even if, for instance, production is efficient because the firm's yield is a high percentage of its capacity, production may not be effective—in terms of market share—if the firm must hold back products owing to poor market conditions (van Mieghem and Dada 1999). Finally, the "equity" criterion concerns equity value performance from the perspective of shareholders. Equity concerns are relevant to board members who must decide whether (or not) a risk management program can be implemented (Cohen and Kunreuther 2007).

In the integrated risk management of an MNC, a crucial question is whether the operations–finance optimization should be effected by adopting a centralized or rather a decentralized approach (see Sect. 2.1). From the perspective of organizational economics, Table 4.2 lists and details five dimensions pertaining to the choice of approach when seeking to integrate operations and finance. We address each of these dimensions in turn.

Table 4.2 Aspects of optimizing the integration of MNC operations and finance

Dimension	Structure	Activity	Trade-offs	References
Organization structure	Decentralization/ coordination	Operations: capacity, production, allocation; Finance: accounting, treasury, capital	Environmental, industry- and firm-specific factors	Martinez and Jarillo (1989)
Objective setting	Decentralization/ coordination	Operations: (i) firm value maximization; (ii) risk reduction via real options Finance: (i) equity value maximization; (ii) risk sharing via financial derivatives	Functional, service, trading orientations	Lessard and Zaheer (1996)
Incentives alignment	Decentralization/ coordination	Operating profit maximization; cost reduction; loss aversion; coordination of incentives	Synergistic benefits versus coordination cost	Kleindorfer and Saad (2005)
Knowledge management	Centralization	Cross-divisional knowledge transfer (technological standardization); cross-functional knowledge diffusion (trans-specialist training)	Specialization benefits versus knowledge distribution cost	Kogut and Zander (1993)
Information management	Centralization	Internal audit system; internal consulting/ informal communication	Integration benefits versus infrastructure/ sharing cost	Rai et al. (2006)

(1) In terms of *organization structure*, most MNCs have decentralized their operations and finance divisions but retain coordination between the two functions. Two types of coordination mechanisms are evident: formal and informal approaches. The formal mechanisms incorporate (i) departmentalization by grouping activities into functional units; (ii) standardization by establishing formal regulations, procedures, and job descriptions; (iii) joint planning—such as capital budgeting as well as goal and strategy setting—to guide the activities of various departments; and (iv) performance control, which includes not only output control (based on filing of reports) but also behavior control (via direct personal surveillance). The informal approaches include (i) lateral relations, or direct contact and task sharing among managers in various functional units; (ii) informal communication by information sharing via personal contact across divisions; and (iii) corporate culture development based on the company's value and vision (Martinez and Jarillo 1989).

(2) As for the *objective setting*, the MNC's operations units typically focus on increasing firm value or maximizing expected profits while the finance department aims to maximize equity value. Operational flexibility seeks to reduce risk by taking positions in real options, and financial hedging shares risk via derivatives contracts with financial institutions. An MNC's divisions of finance, accounting, and treasury are subject to various sociocognitive interpretations concerning their roles when interacting with operational departments. Three types of orientation can be identified: (i) a "service" orientation, whereby financial departments play a supporting role with respect to operational divisions; (ii) a "functional" orientation, as when the firm's treasury department focuses on financial risk exposure when managing revenue, cash flow, and the balance sheet; and (iii) a "trading" orientation that targets long-term value enhancement by mitigating risks. All three of these orientations merit attention, since promoting one at the expense of others could well result in inefficiency. For example, a predominantly functional orientation may identify financial exposure yet fail to acknowledge the competitive exposure that could result from accounting ambiguities (Lessard and Zaheer 1996).

(3) The *alignment of incentives* is required to facilitate cross-functional cooperation in the areas of value-based management and risk mitigation. Coordination incentives should prevail across an MNC's functions and also across supply chain partners; the reason is that integrated risk management can be optimized only through the joint cooperation of various entities. For instance, treasury and operational departments both should be involved in using performance measurement to steer the strategic hedging of competitive exposure. Operations and finance should likewise both contribute to maximizing profit or utility, and to minimizing costs, based on the firm's risk attitude and objective formulation (Kleindorfer and Saad 2005).

(4) In *knowledge management*, centralized knowledge transfer and sharing is crucial for various divisions in a firm and for the partners that collaborate in a supply chain. Such transfer is indispensable for ensuring technological standardization across divisions or companies within a product's life

cycle—especially in knowledge-intensive sectors such as the pharmaceutical industry. Furthermore, knowledge diffusion is necessary for successful trans-specialist training across functional units. In this sense, firms can be viewed as communities that create knowledge and then translate it into value-enhancing products and services (Kogut and Zander 1993).

(5) *Information management* seeks to ensure both data consistency and the high-velocity transfer of strategic and operational information across functional and supply chain entities. Data consistency is the extent to which data definitions and information sharing have been established along the firm's supply chain. As articulated in our closed-loop view, integrated risk management aims to synchronize the physical, financial, and information flows within a firm and across supply chain partners. The key to linking material and monetary flow activities is establishing an efficient information system for real-time communication among various functions. It follows that a supply chain–related competitive advantage is determined by the firm's capacity to process information for the benefit of internal audits and of managing supplier and customer relationships (Rai et al. 2006).

In addition to the perspective of a valuation mechanism for financial instruments (in Sect. 2.3) and the process perspective on operational strategies (Sect. 3.3), we offer here an alternative classification based on risk treatment. Table 4.3 provides details on operational strategies and financial instruments based on four categories of such treatment.

(1) *Risk taking* refers to the base case where firms do not employ any risk mitigation strategy and instead rely on a single particular scenario transpiring. For instance, if a manufacturing firm produces only in foreign locations or sources only from foreign suppliers, then its costs will be denominated in foreign currency and so its profits will be exposed to exchange rate risk. And if a firm declines to adopt any financial hedging instruments and simply assumes that a financial metric will move in only one direction, then its cash flow will be at the mercy of any reversals in that metric. Risk taking can be viewed as speculation, which exposes firms to greater economical and transactional risks.
(2) *Risk sharing* is an action that transfers, via contract, the firm's risk exposure to another party. For example, currency risk sharing allows a supplier to transfer an agreed-upon portion of their transaction's exchange rate exposure to the buyer. In addition, financial derivatives contracts (forwards, futures, options, swaps, etc.) transfer financial index exposure from the derivative's buyer (the firm) to its seller (a financial institution).
(3) *Risk reduction* is any strategy or action intended to reduce the likelihood or consequences of exposure to risk. A good example of risk reduction is the manufacturing firm that operates in both domestic and foreign markets; this is known as "operational diversification" or a "natural hedge". Firms can also shift production volumes between plants located in various currency zones,

Table 4.3 Classification of risk treatment strategies and instruments

Risk treatment	Operational strategy	Financial instrument	References
Risk taking (speculation)	Betting on a single scenario—e.g., full foreign production (economic exposure)	Betting on financial index movement (e.g., exchange rate) by accurate forecast based on financial expertise/valuable information (economic exposure)	Stulz (1996), Shapiro (2002)
	Sourcing from a single supplier—e.g., full foreign procurement (transactional exposure)		
Risk sharing (transfer)	Sharing gains/losses on a financial asset (e.g., exchange rate) with supplier or buyer by (e.g., currency) risk-sharing contract (transactional exposure)	Hedging financial risk by trading derivatives—including forwards, futures, options, swaps—on (e.g., exchange or interest) rates (economic exposure)	Kouvelis (1999), Ding et al. (2007)
	Transferring financial risk (e.g., currency exposure) to supplier or customer by (e.g., exchange rate) pass-through in pricing decisions (transactional exposure)		
Risk reduction (diversification/flexibility)	Designing a dispersed global manufacturing network, also called "operational diversification" or "natural hedge" (economic exposure)	Diversifying financial assets across nations for more stable return and more diffuse risk (economic exposure)	Kogut and Kulatilaka (1994), Shapiro (2002), van Mieghem (2012)
	Shifting production among plants in different currency zones to take advantage of various scenarios, also called "operational flexibility" (economic exposure)		
Risk avoidance (elimination)	Pricing the supply contract payment in domestic/base currency—e.g., US dollars (transactional exposure)	Establishing a monetary reserve in supplier/foreign currency for payment (transactional exposure)	Carter and Vickery (1988), Chowdhry and Howe (1999), Simchi-Levi et al. (2003)
	Integrating foreign supply resources via merger and acquisition (transactional exposure)		

which is referred to as "production flexibility" (see Table 3.1). Analogously, financial diversification reduces the economic exposure of a financial portfolio.

(4) *Risk avoidance* applies when sources of uncertainty are taken under permanent control or when an organization's risk exposure is eliminated altogether. For instance, a firm whose sourcing contracts are denominated in its own (domestic) currency has eliminated its exposure to currency risk. Exchange rate risk can be avoided also if the buyer reserves foreign currency for purchasing supplies. Yet another example: once a buyer acquires a supplier, there is no longer any transactional exposure.

Strategies for sharing, reducing, and avoiding risk are proactive: in fair weather, prepare for foul. "An ounce of prevention is worth a pound of cure" and so proactive risk management is far preferable to coping, after the fact, with the negative outcomes of assuming risks. At the same time, strategies for dealing with risk—such as contingency plans and real-time crisis management—are crucial given the increasing unpredictability and ambiguity of risks. Although fundamental solutions are better than temporary ones, this intuition may not be valid with regard to managing a risk (e.g., exchange rate uncertainty) that is "double edged". In such cases, risk avoidance reduces the downside hazard but also the upside potential; hence the optimal extent of risk reduction depends strongly on the particular circumstances. For strictly downside risks, avoidance is always preferable to either risk reduction or risk sharing.

4.4 Integrated Operational and Financial Hedging

So far we have discussed qualitatively the conditions, process, and dimensions of integrating operations and finance. In order to quantify the value of joint operational and financial hedging, we now present numerical examples in currency and commodity risk management; these examples are adapted from Sodhi and Tang (2012) and van Mieghem (2012).

Example 4.1 A global manufacturer has production plants located in China and the United States to fulfill demand in those respective markets. Both product demand and the exchange rates are uncertain. The firm wonders whether it should hedge financially or operationally. The unit market price is c dollars in the United States and d yuan in China; the unit production cost is e dollars in the United States and f yuan in China. Suppose that demand states are correlated with exchange rates, so that the greater is a country's market demand, the more valuable is its currency. Then there are two equally likely scenarios:

1. US demand is a units, Chinese demand is b units, $a > b$, and the exchange rate is $(\alpha - \beta)$ dollars/yuan;
2. US demand is b units, Chinese demand is a units, $a > b$, and the exchange rate is $(\alpha + \beta)$ dollars/yuan.

Option I: Natural hedging. The global firm produces and sells in local markets. Therefore, the operating profits in the two scenarios are $\pi_1^{\mathrm{I}} = a(c-e) + b(d-f)(\alpha-\beta)$ and $\pi_2^{\mathrm{I}} = b(c-e) + a(d-f)(\alpha+\beta)$.
Option II: Natural and financial hedging. The global firm combines natural hedging with financial hedging, where the latter consists of selling h units of future yuan for α dollars per yuan. The expected payoff from financial hedging is zero, and we ignore the small transaction costs. Then the operating profits in the two scenarios are $\pi_1^{\mathrm{II}} = a(c-e) + b(d-f)(\alpha-\beta) + h\beta$ and $\pi_2^{\mathrm{II}} = b(c-e) + a(d-f)(\alpha+\beta) - h\beta$.
Option III: Operational hedging. The global firm adopts operational hedging via allocation flexibility; hence it produces only in China (scenario 1) or only in the United States (scenario 2). Here the operating profits in the two scenarios are $\pi_1^{\mathrm{III}} = ac + bd(\alpha-\beta) - f(a+b)(\alpha-\beta)$ and $\pi_2^{\mathrm{III}} = bc + ad(\alpha+\beta) - e(a+b)$.
Option IV: Operational and financial hedging. The global firm combines allocation flexibility with financial hedging. In this case, the operating profits in the two scenarios are $\pi_1^{\mathrm{IV}} = ac + bd(\alpha-\beta) - f(a+b)(\alpha-\beta) + h\beta$ and $\pi_2^{\mathrm{IV}} = ac + bd(\alpha-\beta) - e(a+b) - h\beta$.
Option V: Operational and perfect financial hedging. The global firm combines allocation flexibility with *perfect* financial hedging—that is, hedging that achieves zero variance. Now the operating profits in the two scenarios are $\pi_1^{\mathrm{IV}} = ac + bd(\alpha-\beta) - f(a+b)(\alpha-\beta) + h_p\beta$ and $\pi_2^{\mathrm{IV}} = ac + bd(\alpha+\beta) - e(a+b) - h_p\beta$.

In Example 4.1, we set the following parameter values: a = 2,000,000, b = 1,000,000, c = 3000, d = 20,000, e = 1500, f = 10,000, $\alpha = 0.15$, $\beta = 0.05$, $h = 10\,\mathrm{kM}$, and $h_p = 15\,\mathrm{kM}$. Then we can write $\pi_1^{\mathrm{I}} = 2\,\mathrm{M} \times 1.5\,\mathrm{k} + 1\,\mathrm{M} \times 10\,\mathrm{k} \times 0.1 =$ (\$)4 kM and $\pi_2^{\mathrm{I}} = 1\,\mathrm{M} \times 1.5\,\mathrm{k} + 2\,\mathrm{M} \times 10\,\mathrm{k} \times 0.2 =$ (\$5.5 kM). The expected profit of Option I is $\mathbb{E}[\pi^{\mathrm{I}}] = 4.75\,\mathrm{kM}$, and its standard deviation is $\sigma^{\mathrm{I}} = 0.75\,\mathrm{kM}$. In Option II, we have $\pi_1^{\mathrm{II}} = 4.5\,\mathrm{kM}$, $\pi_2^{\mathrm{II}} = 5\,\mathrm{kM}$, $\mathbb{E}[\pi^{\mathrm{II}}] = 4.75\,\mathrm{kM}$, and $\sigma^{\mathrm{II}} = 0.25\mathrm{kM}$. In Option III, $\pi_1^{\mathrm{III}} = 2\,\mathrm{M} \times 3\,\mathrm{k} + 1\,\mathrm{M} \times 20\,\mathrm{k} \times 0.1 - 3\,\mathrm{M} \times 10\,\mathrm{k} \times 0.1 = 5\,\mathrm{kM}$, $\pi_2^{\mathrm{III}} = 1\,\mathrm{M} \times 3\,\mathrm{k} + 2\,\mathrm{M} \times 20\,\mathrm{k} \times 0.2 - 3\,\mathrm{M} \times 1.5\,\mathrm{k} = 6.5\,\mathrm{kM}$, $\mathbb{E}[\pi^{\mathrm{III}}] = 5.75\,\mathrm{kM}$, and $\sigma^{\mathrm{III}} = 0.75\,\mathrm{kM}$. In Option IV, $\pi_1^{\mathrm{IV}} = 5.5\,\mathrm{kM}$, $\pi_2^{\mathrm{IV}} = 6\,\mathrm{kM}$, $\mathbb{E}[\pi^{\mathrm{IV}}] = 5.75\,\mathrm{kM}$, and $\sigma^{\mathrm{IV}} = 0.25\,\mathrm{kM}$. Finally, in Option V we have $\pi_1^{\mathrm{V}} = 5.75\,\mathrm{kM}$, $\pi_2^{\mathrm{V}} = 5.75\,\mathrm{kM}$, $\mathbb{E}[\pi^{\mathrm{V}}] = 5.75\,\mathrm{kM}$, and $\sigma^{\mathrm{V}} = 0$. The various options are illustrated in Fig. 4.8.

To illustrate the value of integrated operational and financial hedging, we conduct a mean–standard deviation analysis of the five options, where the expected profit represents return and the standard deviation represents risk. The firm's objective function can be expressed as $u = \mathbb{E}[\pi] - \lambda\sigma$, where $\lambda = 0.1$ is the coefficient of risk aversion. Hence the utilities of the five options are $u^{\mathrm{I}} = 4.675\,\mathrm{kM}$, $u^{\mathrm{II}} = 4.725\,\mathrm{kM}$, $u^{\mathrm{III}} = 5.675\,\mathrm{kM}$, $u^{\mathrm{IV}} = 5.725\,\mathrm{kM}$, and $u^{\mathrm{V}} = 5.75\,\mathrm{kM}$. Moving from Option I to Option II via financial hedging increases the firm's utility by reducing risk without affecting expected profit, whereas moving from Option I to

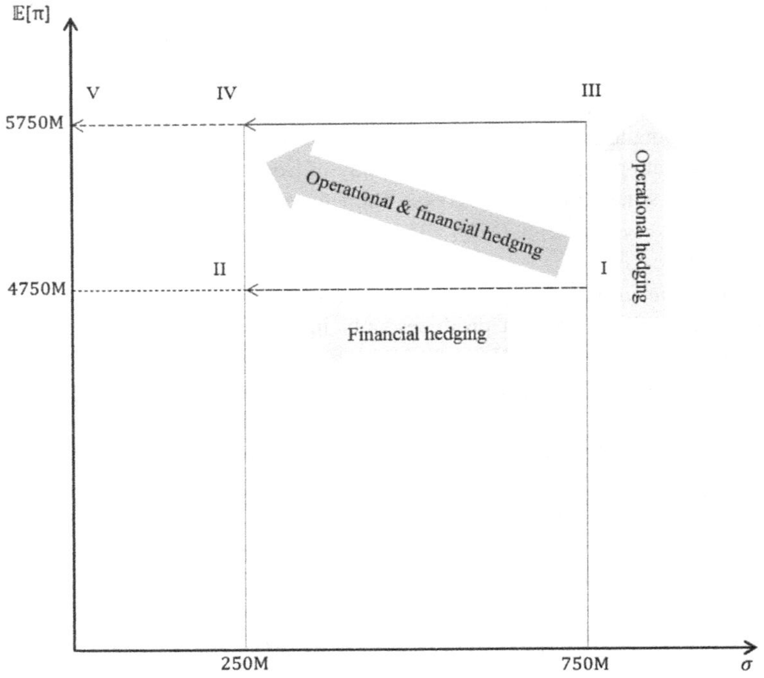

Fig. 4.8 Effect of operational and financial hedging

Option III via operational hedging increases utility by increasing expected profit yet *without* mitigating the risk of variability. The integration of operational and financial hedging in Options IV and V increases return and also reduces risk, and the optimal value of firm utility is achieved under financial hedging with zero variance.

Example 4.2 An agriculture company in Europe plants and sells corn, and it faces two equally probable scenarios of annual yield:

1. when the firm's yield is high ($4a$ bushels), the market price is €$2b$;
2. when the firm's yield is low ($2a$ bushels), the market price is €$4b$.

Option I: No hedging. If the firm does not take any action, then the revenue in the two scenarios are the same: $R_1^{\mathrm{I}} = R_2^{\mathrm{I}} = 8ab$. Therefore, expected revenue is $\mathbb{E}[R^{\mathrm{I}}] = (€)8ab$ and the standard deviation is $\sigma^{\mathrm{I}} = 0$.
Option II: Financial hedging. If the agriculture company adopts financial hedging via commodity derivatives—selling $3a$ bushels of corn in a futures contract at the price of €$3b$ per bushel—then the firm's revenue in scenario 1 is $R_1^{\mathrm{I}} = 3a \times$

$3b + a \times 2b = 11ab$ and in scenario 2 is $R_2^{\text{II}} = 3a \times 3b - a \times 4b = 5ab$. Hence expected revenue is $\mathbb{E}[R^{\text{II}}] = (€)8ab$ and the standard deviation is $\sigma^{\text{II}} = 3ab$. Thus, financial hedging via futures contracts increases variability risk without affecting expected revenue.

Option III: Integrated financial and operational hedging. If the firm adopts operational hedging by storing a bushels of corn in scenario 1 and sells them in scenario 2—and if we ignore spoilage and the holding costs of inventory—then the firm's revenue in the two scenarios are the same: $R_1^{\text{III}} = R_2^{\text{III}} = (\ €\ )9ab$. It follows that expected revenue is $\mathbb{E}[R^{\text{III}}] = (\ €\ )9ab$ and the standard deviation is $\sigma^{\text{III}} = 0$. Thus integrated operational and financial hedging leads not only to higher expected revenue and but also to lower risk of variability.

Finally, we conduct a mean–standard deviation analysis of these three options. If the focal firm's objective function is the same as in Example 4.1 (i.e., $u = \mathbb{E}[\pi] - \lambda\sigma$ with $\lambda = 0.1$), then the utilities of Example 4.2's three options are $u^{\text{I}} = 8ab$, $u^{\text{II}} = 7.7ab$, and $u^{\text{III}} = 9ab$.

References

Barney, J. B. (1991). Firm resources and sustained competitive advantage. *Journal of Management, 17,* 99–121.

Carter, J. R., & Vickery, S. K. (1988). Managing volatile exchange rates in international purchasing. *Journal of Purchasing Materials Management, 24,* 13–20.

Chapman, R. J. (2006). *Simple tools and techniques for enterprise risk management.* New Jersey, U.S.A.: Wiley.

Chowdhry, B., & Howe, J. T. B. (1999). Corporate risk management for multinational corporations: Financial and operational hedging policies. *European Finance Review, 2,* 229–246.

Cohen, M. A., & Kunreuther, H. (2007). Operations risk management: Overview of Paul Kleindorfer's contributions. *Production and Operations Management, 16,* 525–541.

Ding, Q., Dong, L., & Kouvelis, P. (2007). On the integration of production and financial hedging decisions in global markets. *Operations Research, 55,* 470–489.

Ghoshal, S., & Nohria, N. (1993). Horses for courses: Organizational forms for multinational corporations. *Sloan Management Review, 34,* 23–35.

Kleindorfer, P. R., & Saad, G. H. (2005). Managing disruption risks in supply chain. *Production and Operations Management, 14,* 53–68.

Kogut, B., & Kulatilaka, N. (1994). Operating flexibility, global manufacturing, and the option value of a multinational network. *Management Science, 40,* 123–139.

Kogut, B., & Zander, U. (1993). Knowledge of the firm and the evolutionary theory of the multinational enterprise. *Journal of International Business Studies, 24,* 625–645.

Kouvelis, P. (1999). Global sourcing strategies under exchange rate uncertainty. In S. Tayur, R. Ganeshan, & M. Magazine (Eds.), *Quantitative models for supply chain management* (pp. 625–668).Dordrecht: Kluwer Academic Publishers.

Lessard, D. R., & Zaheer, S. (1996). Breaking the silos: Distributed knowledge and strategic responses to volatile exchange rates. *Strategic Management Journal, 17,* 513–533.

Martinez, J. I., & Jarillo, J. C. (1989). The evolution of research on coordination mechanisms in multinational corporations. *Journal of International Business Studies, 20,* 489–514.

Miller, K. (1992). A framework for integrated risk management in international business. *Journal of International Business Studies, 23,* 311–331.

Nawrocki, D. (1999). A brief history of downside risk measures. *Journal of Investing, 8*(3), 9–26.

Otley, D. (1999). Performance management: A framework for management control system design. *Management Accounting Research, 10,* 363–382.

Porter, M. E. (1985). *Competitive advantage: Creating and sustaining superior performance.* New York: The Free Press.

Rai, A., Patnayakuni, R., & Seth, N. (2006). Firm performance impacts of digitally enabled supply chain integration capabilities. *MIS Quarterly, 30,* 225–246.

Schmieder-Ramirez, J., & Mallette, L. (2015). Using the SPELIT analysis technique for organizational transitions. In M. Carmo (Eds.), *Education applications and developments* (pp. 291–300). Science Press.

Shapiro, A. (2002). International Financial Management. John. Wiley & Sons, London.

Simchi-Levi, D., Kaminsky, P., & Simchi-Levi, E. (2003). *Designing and managing the supply chain*. New York: McGraw-Hill/Irwin.

Sodhi, M., & Tang, C. (2012). Tactical approaches for mitigating supply chain risks: Financial and operational hedging. In M. Sodhi & C. Tang (Eds.) *Managing supply chain risk* (pp. 109–133). Berlin: Springer.

Stulz, R. (1996). Rethinking risk management. *Journal of Applied Corporate Finance, 9,* 8–25.

The Committee of Sponsoring Organizations of the Treadway Commission [COSO]. (2004). Enterprise risk management integrated framework executive summary, September 2004.

van Mieghem, J. A. (2003). Capacity management, investment and hedging: Review and recent developments. *Manufacturing & Service Operations Management, 5,* 269–302.

van Mieghem, J. A. (2012). Risk management and operational hedging: An overview. In P. Kouvelis, L. Dong, O. Boyabatli, & R. Li (Eds.), *Handbook of integrated risk management in global supply chains* (pp. 13–49). New York: Wiley.

van Mieghem, J. A., & Dada, M. (1999). Price versus production postponement: Capacity and competition. *Management Science, 45,* 1631–1649.

Wernerfelt, B. (1984). A resource-based view of the firm. *Strategic Management Journal, 5,* 171–180.

Chapter 5
Integrated Risk Management with Capacity Reshoring

In this chapter, we provide an analytical model of integrated risk management with capacity reshoring. First, we review the analytical models on integrated operations-finance risk management in global manufacturing. Second, we formulate a global supply chain model and propose an optimal integration of operational flexibility and financial hedging. Third, we study analytically and numerically the interactions between operations and finance. Furthermore, the main insights and suggested directions for future research are summarized.

5.1 Integrated Operational Flexibility and Financial Hedging

Multinational corporations (MNCs) operating in global markets are typically exposed to foreign exchange volatility, supply uncertainty, and demand risk (*The Economist* 2009). The rising costs and reliability issues in emerging economies, such as China and Mexico, have led to a rapid growth in firms' *reshoring* in order to re-establish reliable domestic capacity (de Treville and Trigeorgis 2010; Cohen and Lee 2015). The Reshoring Initiative, a nonprofit advocacy group, reports that major US-based companies—including Ford, General Electric, and GM—are relocating production domestically (Sauter and Stebbins 2016); these actions returned approximately 265,000 jobs to the United States from January 2010 through June 2016 (Selko 2016). The sportswear group Adidas has similarly decided to bring a portion of its shoe production back to Germany by building a highly automated factory near Ansbach that enables both fulfillment of excess demand and the late customization of products with fast delivery (*Financial Times* 2016). Such capacity

---

This chapter is a revised version of "Integrated operational and financial hedging with capacity reshoring" published in issue 260 of *European Journal of Operations Research*.

L. Zhao and A. Huchzermeier, *Supply Chain Finance*, EURO Advanced Tutorials on Operational Research, https://doi.org/10.1007/978-3-319-76663-8_5

reshoring typically leads to a production network consisting of more domestic facilities and limited foreign capacity. For instance, the automotive manufacturer Daimler now produces its C-Class Sedan not only in Beijing, China, but also in Bremen and Sindelfingen, Germany (Daimler 2014). The Sindelfingen plant is geologically stable and therefore more reliable than the Bremen plant, which is located along the North Sea and thus is exposed to flooding risk under climate change (European Commission 2009). The Beijing plant is similarly vulnerable to, for instance, devastating summer rainstorms (*The Economist* 2012).

In addition to bolstering their reliable domestic production,[1] multinational corporations often execute switching options that shift production across different countries (Kogut and Kulatilaka 1994; Huchzermeier and Cohen 1996; Kazaz et al. 2005) in order to manage supply–demand mismatches and currency risk. Moreover, global firms such as Daimler, Ford, and GM typically use financial derivatives to reduce the impact of currency rate fluctuations (Daimler 2016; Ford 2016; GM 2016). The integration of production switching and financial hedging is a topic frequently explored in the literature (Mello et al. 1995; Chowdhry and Howe 1999; Hommel 2003; Ding et al. 2007; Chen et al. 2014). However, there is a dearth of research that analyzes *capacity reshoring* and its interaction with production switching and financial hedging—despite their prevalence in practice.

The extant literature has examined integrated production switching and financial hedging in a risk-neutral context (Mello et al. 1995) or through the lens of models that use variance-based risk measures (Chowdhry and Howe 1999; Hommel 2003; Ding et al. 2007; Chen et al. 2014). Yet such measures, including those employed by mean-variance (MV) models, treat upside potential and downside risk equally. As a downside risk measure, value-at-risk (VaR) focuses on the probability—not the magnitude—of risk (Acerbi and Tasche 2002). Conditional value-at-risk (CVaR) remedies VaR by taking both risk probability and magnitude into account; the resulting risk measure is *coherent* in the sense that it fulfills the axioms of convexity, monotonicity, subadditivity, translation equivariance, and positive homogeneity (Artzner et al. 1999; Choi et al. 2011). We therefore adopt CVaR as a coherent risk measure. In practice, a CVaR constraint can be applied in three cases: when internal capital is required for an investment opportunity (Froot et al. 1993), when a lower bound of profitability has been imposed by shareholders (Hommel 2003), or when a standing loan is tied to default risk (Gamba and Triantis 2014). Although CVaR has been adopted in operations (e.g., Tomlin and Wang 2005; Gotoh and Takano 2007), mean-CVaR has received little attention in integrated operational and financial hedging despite its popularity in finance (e.g., Chen et al. 2012; Iyengar and Ma 2013).

In sum, this chapter attempts to fill these gaps in the literature by addressing three research questions, as follows.

---

[1]In this chapter, we use the terms "reliable domestic production" and "capacity reshoring" interchangeably.

(i) Which risk management tools can mitigate currency and mismatch risks more effectively?
(ii) How can a mean-CVaR framework be implemented in a MNC with decentralized functions?
(iii) What are the relationships among capacity reshoring, production switching, and financial hedging under mean-CVaR?

To answer these questions, we examine a global manufacturer that adopts capacity reshoring, production switching, and financial hedging to manage supply–demand mismatches and currency risk. Applying the mean-CVaR optimization proposed by Krokhmal et al. (2002), we decompose operations and finance in a MNC in this way: operational departments maximize expected profit subject to a CVaR constraint, while finance departments focus on financial hedging to minimize CVaR subject to a minimum expected profit. Our first result is that the complementarity between operational flexibility and financial hedging is due mainly to profitability enhancement. Operational flexibility drives profitability and reduces downside risk, while financial hedging minimizes downside risk and can alter feasible set of capacity portfolios indirectly through the CVaR constraint. Second, operational flexibility and financial hedging are substitutes in terms of risk reduction. With operational flexibility, financial options hedge only rare and extreme exchange rates because real options are expected to generate greater profits. Financial hedging is more effective at reducing risk in CVaR when it is *not* used in conjunction with operational flexibility, while capacity reshoring and production switching are complements (resp., substitutes) when used to fulfill foreign (resp., domestic) demand. Third, operations and finance departments should collaborate so as to minimize their substitution effects. The efficiency of financial hedging depends on accurate estimates of cash flow distribution as shaped by operational flexibility; at the same time, the CVaR constraint dictates that the feasible set of capacity portfolios be determined by financial hedging.

5.2 Global Production with Capacity Reshoring, Switching Options, and Financial Hedging

This section presents our model formulation, timeline, and objective function. We consider a global firm that mitigates the risks of random capacity, uncertain demand, and volatile exchange rate by marshaling three strategies: reliable domestic production, production switching options, and financial hedging via currency derivatives.

Figure 5.1 depicts the global firm's production network. The MNC supplies one product with two domestic capacities K_1 and K_r and one foreign capacity K_2. Currency zone 1 represents the domestic area; currency zone 2 represents the foreign area. The MNC first decides on capacity order quantities $\boldsymbol{Q} = (Q_1, Q_2, Q_r)^{\mathrm{T}} \in \mathbb{R}_+^3$ at unit capacity cost c_i^k for resource $i = 1, 2, r$.

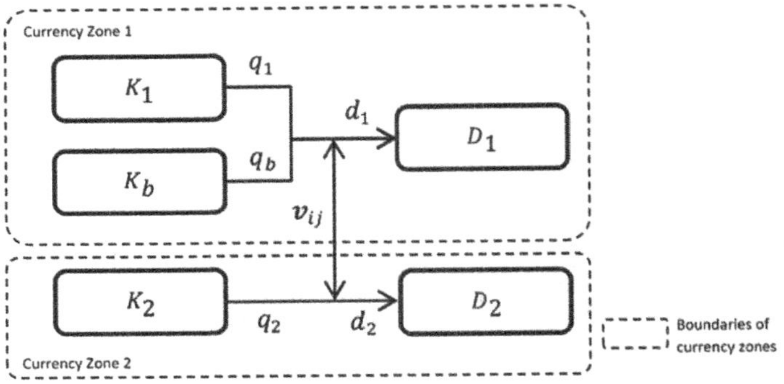

Fig. 5.1 The global production network

Throughout the chapter, we use boldface letters to denote vectors and superscript T to signify transposition. Capacity K_i can be unreliable and subject to the proportional random yield factor $R_i \in (0, 1]$ for $i = 1, 2, r$. All capacities have endogenous production function $K_i = Q_i R_i$ (cf. Dada et al. 2007) for $i = 1, 2, r$, so the global firm might receive only a fraction of the capacity quantity ordered. Domestic capacity K_r is perfectly reliable (i.e., $R_r = 1$); therefore, $K_r = Q_r$. Table 5.1 summarizes our notation and assumptions.

Market demand D_i in the currency zone $i = 1, 2$ is stochastic with probability density function (PDF) $g_i(\cdot)$. The foreign currency exchange rate $s = \frac{\text{domestic currency}}{\text{foreign currency}}$ for $s \in [\underline{s}, \overline{s}]$ has a PDF of $e(\cdot)$ that is assumed to be independent of demand rates D_i. The joint distribution of exchange rate and supply–demand matching states (i.e., s, K_i, D_i) has density function $f(\cdot)$ with support Ω_i (i = 1, 2, …,10). We assume that the global firm is able to estimate the distribution of uncertainty factors, which means that the probabilities P_i of the respective states Ω_i are known ex ante. Once capacity has been established, the global firm decides on production quantities q_i at unit production cost c_i for $i = 1, 2, r$. We assume that $c_1^k < c_r^k$ and $c_1 < c_r$, given that *reliable* domestic capacity and production are more expensive than *ordinary* domestic capacity and production (Tomlin 2006). The global firm has switching options to shift v_{ij} production units from currency zone i to j for $i, j = 1, 2$ and $i \neq j$; however, the execution of such an option incurs a unit switching cost c_{ij}^s (in the destination zone's currency). The MNC then decides on its output quantity d_i destined for market $i = 1, 2$. The selling price p_i in market $i = 1, 2$ is assumed to be exogenous; also, the conditions $p_1 > c_r > c_1$ and $p_2 > c_2$ ensure the profitability of unit production in each market.

Figure 5.2 illustrates our two-stage stochastic program for characterizing the MNC's operational and financial decisions. In Stage 1, the global firm invests in domestic and foreign capacities $\boldsymbol{Q}$ subject to a CVaR constraint. The MNC purchases financial hedging contracts of forwards or put options on foreign currency,

Table 5.1 Summary of notation and assumptions

Symbol	Description	Assumptions
s	Exchange rate in currency zone 2	$s = \frac{\text{domestic currency}}{\text{foreign currency}}$, $s \in [\underline{s}, \bar{s}]$
$e(\cdot)$	Probability density function (PDF) of exchange rate distribution	Exogenous, stochastic; independent of D_i
K_i	Ordinary capacity in currency zone i for $i = 1, 2$	Unreliable with endogenous production function; $K_i = Q_i R_i$ for $i = 1, 2$
K_r	Reliable (domestic) capacity in currency zone 1	Perfectly reliable; $K_r = Q_r$
R_i	Proportional random yield factor of capacity K_i for $i = 1, 2$	Random variables; $R_i \in (0, 1]$
Q_i	Capacity order quantity at resource i for $i = 1, 2, r$	Decision variable in stage 1; $\boldsymbol{Q} \in \mathbb{R}^3_+$
q_i	Production quantity at resource i for $i = 1, 2, r$	Decision variable in stage 2
c_i^k	Unit capacity cost at resource i for $i = 1, 2, r$	$c_j^k < c_r^k$ for $j = 1, 2$
c_i	Unit production cost at resource i for $i = 1, 2, r$	$c_j < c_r$ for $j = 1, 2$
v_{ij}	Production quantity switched from currency zone i to j	$i, j = 1, 2,\ i \neq j$
c_{ij}^s	Unit switching cost of shifting a production unit from currency zone i to j in the destination location's (j's) currency	Exogenous, deterministic $i, j = 1, 2,\ i \neq j$
D_i	Market demand in currency zone i for $i = 1, 2$	Exogenous, stochastic
$g_i(\cdot)$	PDF of demand distribution for market i for $i = 1, 2$	Exogenous; known in stage 1
y	Random vector $\boldsymbol{y} \triangleq (\boldsymbol{K}, s, \boldsymbol{D})$	$\boldsymbol{y} \in \mathbb{R}^m$
$f(\cdot)$	Density function of joint distribution of exchange rate and supply–demand matching states	Exogenous, has support $\Omega_i (i = 1, 2, \ldots, 10)$
P_i	Probability of joint exchange rate and supply–demand matching state Ω_i	Exogenous, known in stage 1
d_i	Product output in currency zone i for $i = 1, 2$	Decision variable in stage 2; $d_i \leq D_i$
p_i	Market (unit) price in currency zone i for $i = 1, 2$	Exogenous; $p_i > c_r > c_i$ for $j = 1, 2$
$H(h)$	Payoff of financial hedging portfolio	$\mathbb{E}[H(\boldsymbol{h})] = 0$
h_i	Size vector of currency forwards or options, $i = f, p$	Consists of currency forwards or options
$C(h)$	Overall cost of financial hedging	Includes commission costs, transaction fees, and margin requirements
α	Threshold value of CVaR	Endogenous; depends on β
β	Confidence interval of CVaR	Endogenous, deterministic
λ	Coefficient of risk aversion in mean-CVaR objective	$\lambda \geq 0$

(continued)

Table 5.1 (continued)

Symbol	Description	Assumptions
ω	CVaR constraint on operations	Endogenous, deterministic
ρ	Minimum expected profit constraint	Endogenous, deterministic

and $\boldsymbol{h}_i \in \mathbb{R}^n$ $(i = f, p)$[2] is the size vector of currency forwards or options. The firm has postponement flexibility (van Mieghem and Dada 1999) when planning production or switching activities that are contingent on the realization of uncertainty. This last assumption alludes to the capacity or key component commitment observed in practice.

In Stage 2—after observing realized capacities $\boldsymbol{K} = (K_1, K_r, K_2)$, demands $\boldsymbol{D} = (D_1, D_2)$, and the exchange rate s—the global firm decides on production quantities $\boldsymbol{q} = (q_1, q_r, q_2)$ and switching quantities $\boldsymbol{v}_{ij} = (v_{12}, v_{21})$. The firm first decides on regular production quantities and then on the execution of reliable production and switching options to match supply with demand. Note that the expected payoff from financial hedging is assumed to be zero (this is the *no-arbitrage* condition): $\mathbb{E}[H(\boldsymbol{h})] = 0$, where $\mathbb{E}[\cdot]$ is the expectation operator. In the absence of arbitrage, the overall hedging cost (which includes commission fees, transaction costs, and margin requirements) is equal to the payout from financial hedging: $\mathbb{E}[C(\boldsymbol{h})] = \mathbb{E}[R(\boldsymbol{h})e^{-\gamma t}]$. Thus the MNC's final profit is $\Pi(\boldsymbol{Q}, \boldsymbol{h}, \boldsymbol{y}) = \Pi(\boldsymbol{Q}, \boldsymbol{y}) + R(\boldsymbol{h}, s) - C(\boldsymbol{h})e^{\gamma t}$. Here $\Pi(\boldsymbol{Q}, \boldsymbol{y})$ denotes operating cash flow, and the random vector $\boldsymbol{y} \triangleq (\boldsymbol{K}, s, \boldsymbol{D}) \in \mathbb{R}^m_+$. The payoff from financial hedging, $H(\boldsymbol{h}, s) = R(\boldsymbol{h}, s) - C(\boldsymbol{h})e^{\gamma t}$, reflects both the payout $R(\boldsymbol{h}, s)$ from financial hedging in Stage 2 and the overall hedging cost $C(\boldsymbol{h})e^{\gamma t}$ in Stage 1.[3] The term γ is the risk-free interest rate in domestic currency, and t is the time duration between Stage 1 and Stage 2 as well as the time to maturity of financial derivatives.

To formulate our objective function, we first define CVaR following the approach of Rockafellar and Uryasev (2000). Let $(\Omega; \mathbb{F}; P)$ be a probability space with filtration $\mathbb{F}$ and probability measure P. In our setting, Ω is the probability space on which $f(\cdot)$ is defined and $\mathbb{F}$ is the set of exchange rates and supply–demand state realizations. An uncertain outcome (i.e., loss function) is represented by a measurable function $L : \Omega \to \mathbb{R}$. We specify the vector space $\mathfrak{L}$ of possible functions, where it is sufficient to consider $\mathfrak{L} = \mathcal{L}_\infty(\Omega; \mathbb{F}; P)$. Given the joint operational and financial decisions $(\boldsymbol{Q}, \boldsymbol{h})$, we can write the probability that the loss function $l(\boldsymbol{Q}, \boldsymbol{h}, \boldsymbol{y}) \triangleq -\Pi(\boldsymbol{Q}, \boldsymbol{h}, \boldsymbol{y})$ does not exceed threshold α as $\Psi(\boldsymbol{Q}, \boldsymbol{h}, \alpha) \triangleq \int_{l(\boldsymbol{Q}, \boldsymbol{h}, \boldsymbol{y}) \le \alpha} f(\boldsymbol{y}) \mathrm{d}\boldsymbol{y}$.

---

[2]Here f and p denote (respectively) currency forwards and currency put options. For the sake of brevity, the vector $\boldsymbol{h}_i$ is sometimes simply written as $\boldsymbol{h}$ when no confusion could result.

[3]For brevity, hereafter we use $H(\boldsymbol{h})$ and $R(\boldsymbol{h})$ to signify $R(\boldsymbol{h}, s)$ and $H(\boldsymbol{h}, s)$, respectively.

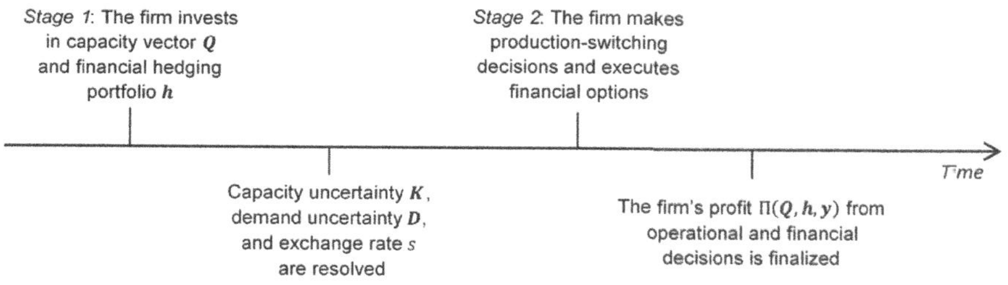

Fig. 5.2 Timeline of events

For a given confidence level $\beta \in (0, 1)$ (e.g., $\beta = 0.9$ or $\beta = 0.95$) and fixed decisions $(\boldsymbol{Q}, \boldsymbol{h})$, the conditional value-at-risk $\mathrm{CVaR}_\beta(\boldsymbol{Q}, \boldsymbol{h}, \boldsymbol{y})$ is defined as the expected value of loss that exceeds value-at-risk $\mathrm{VaR}_\beta(\boldsymbol{Q}, \boldsymbol{h}, \boldsymbol{y}) \triangleq \min\{\alpha \in \mathbb{R}: \ \Psi(\boldsymbol{Q}, \boldsymbol{h}, \alpha) \geq \beta\}$; thus,

$$\mathrm{CVaR}_\beta(\boldsymbol{Q}, \boldsymbol{h}, \boldsymbol{y}) \triangleq \frac{1}{1-\beta} \int_{l(\boldsymbol{Q}, \boldsymbol{h}, \boldsymbol{y}) \geq \mathrm{VaR}_\beta(\boldsymbol{Q}, \boldsymbol{h}, \boldsymbol{y})} l(\boldsymbol{Q}, \boldsymbol{h}, \boldsymbol{y}) f(\boldsymbol{y}) \mathrm{d}\boldsymbol{y}. \tag{5.1}$$

The MNC is risk averse and seeks to maximize the following objective function:

$$\max_{\boldsymbol{Q} \in \mathbb{R}_+^3, \boldsymbol{h} \in \mathbb{R}^n} \{\mathbb{E}[\Pi(\boldsymbol{Q}, \boldsymbol{h}, \boldsymbol{y})] - \lambda \mathrm{CVaR}_\beta(\boldsymbol{Q}, \boldsymbol{h}, \boldsymbol{y})\} \tag{5.2}$$

$$\text{s.t.} \quad \mathrm{CVaR}_\beta(\boldsymbol{Q}, \boldsymbol{h}, \boldsymbol{y}) \leq \omega, \ \mathbb{E}[\Pi(\boldsymbol{Q}, \boldsymbol{h}, \boldsymbol{y})] \geq \rho, \ \mathbb{E}[H(\boldsymbol{h})] = 0$$

(as usual, "s.t." abbreviates "subject to"). Our risk parameter $\lambda \geq 0$ represents the coefficient of risk aversion—that is, the rate at which a firm substitutes CVaR for expected profit.

5.2.1 *Strategy Definition and Mean-CVaR Decomposition*

Next, we define six strategies in terms of their respective degrees of operational flexibility and financial hedging. We then decompose mean-CVaR optimization.

Definition 5.1

(a) *Base case* refers to a nonflexible firm without reliable domestic capacity, production switching, or financial hedging (i.e., $K_r = v_{ij} = \boldsymbol{h} = 0$); ceteris paribus.
(b) *Reliable production only* denotes a partially flexible firm with reliable domestic capacity but without production switching or financial hedging (i.e., $K_r > 0$ and $v_{ij} = \boldsymbol{h} = 0$); ceteris paribus.

(c) *Switching options only* means a partially flexible firm with production switching yet with neither domestic reliable capacity nor financial hedging (i.e., $v_{ij} \geq 0$, and $K_r = \boldsymbol{h} = 0$); ceteris paribus.
(d) *Full operational flexibility* indicates a fully flexible firm with reliable domestic capacity and production switching but without financial hedging (i.e., $K_r > 0$, $v_{ij} \geq 0$, and $\boldsymbol{h} = 0$); ceteris paribus.
(e) *Financial hedging only* refers to a nonflexible firm that adopts financial hedging but employs neither reliable domestic capacity nor production switching (i.e., $\boldsymbol{h} > 0$ and $K_r = v_{ij} = 0$); ceteris paribus.
(f) *Integrated risk management* denotes a fully flexible firm that uses reliable domestic capacity, production switching, and financial hedging (i.e., $K_r > 0$, $v_{ij} \geq 0$, and $\boldsymbol{h} > 0$); ceteris paribus.

Definition 5.1 presents six strategies that vary depending on their respective choices of capacity reshoring, production switching, and financial hedging. Hence we are now in a position to compare the relative effectiveness of various risk management tools. We use $\Pi^x(\boldsymbol{Q},\boldsymbol{h},\boldsymbol{y})$ for $x = \text{bc}, \text{rp}, \text{sw}, \text{op}, \text{fh}, \text{int}$ to denote the profit function when strategy x is adopted[4]; for brevity this function may also be written as $\Pi(\boldsymbol{Q},\boldsymbol{h},\boldsymbol{y})$, $\Pi^x(\boldsymbol{Q},\boldsymbol{h})$, or even Π^x. Similarly, $\text{CVaR}^x_\beta(\boldsymbol{Q},\boldsymbol{h},\boldsymbol{y})$ is defined as the CVaR when $l(\boldsymbol{Q},\boldsymbol{h},\boldsymbol{y}) = -\Pi^x(\boldsymbol{Q},\boldsymbol{h},\boldsymbol{y})$ for strategy $x = \text{bc}, \text{rp}, \text{sw}, \text{op}, \text{fh}, \text{int}$; this term likewise may be denoted also as $\text{CVaR}^x_\beta(\boldsymbol{Q},\boldsymbol{h})$ or CVaR^x_β.

To ensure the tractability of our model, we assume that CVaR is convex (see Rockafellar and Uryasev 2002, Cor. 11), that the expected profit function is concave, and that the decision vector set is (jointly) convex. Following the approach of Krokhmal et al. (2002, p. 49), we show that three different formulations are equivalent to the mean-CVaR optimization.

Lemma 5.1 *Suppose that the functions* $\text{CVaR}_\beta(\boldsymbol{p},\boldsymbol{y})$ *and* $\mathbb{E}[\Pi(\boldsymbol{x},\boldsymbol{y})]$ *depend on the decision vector* $\boldsymbol{p} \triangleq (\boldsymbol{Q},\boldsymbol{h}) \in \mathbf{P}$, *and consider the following three problems*:

$$\max_{\boldsymbol{Q}\in\mathbb{R}^3_+,\boldsymbol{h}\in\mathbb{R}^n} \{\mathbb{E}[\Pi(\boldsymbol{p},\boldsymbol{y})] - \lambda\text{CVaR}_\beta(\boldsymbol{p},\boldsymbol{y})\} \quad \text{s.t.} \quad \boldsymbol{p} \in \mathbf{P} \textit{ and } \lambda \geq 0; \tag{P1}$$

$$\min_{\boldsymbol{h}\in\mathbb{R}^n} \text{CVaR}_\beta(\boldsymbol{p},\boldsymbol{y}) \quad \text{s.t.} \quad \mathbb{E}[\Pi(\boldsymbol{p},\boldsymbol{y})] \geq \rho \textit{ and } \boldsymbol{p} \in \mathbf{P}; \tag{P2}$$

$$\max_{\boldsymbol{Q}\in\mathbb{R}^3_+} \mathbb{E}[\Pi(\boldsymbol{p},\boldsymbol{y})] \quad \text{s.t.} \quad \text{CVaR}_\beta(\boldsymbol{p},\boldsymbol{y}) \leq \omega \textit{ and } \boldsymbol{p} \in \mathbf{P}. \tag{P3}$$

Now suppose that the constraints $\mathbb{E}[\Pi(\boldsymbol{p},\boldsymbol{y})] \geq \rho$ and $\text{CVaR}_\beta(\boldsymbol{p},\boldsymbol{y}) \leq \omega$ *have internal points. Then the efficient frontiers for problems* (P1)–(P3) *can be traced by varying the parameters* λ, ρ, *and* ω *accordingly. When* $\text{CVaR}_\beta(\boldsymbol{p},\boldsymbol{y})$ *is convex,*

[4]We shall use "bc", "rp", "sw", "op", "fh", and "int" to denote (respectively) "base case", "reliable production only", "switching options only", "full operational flexibility", "financial hedging only", and "integrated risk management".

$\mathbb{E}[\Pi(\boldsymbol{p}, \boldsymbol{y})]$ *is concave, and the decision vector set* $\mathbf{P}$ *is convex, then problems* (P1)–(P3) *are equivalent in that they generate the same efficient frontier.*

Proof All proofs are given in the Appendix (available on request).

Lemma 5.1 establishes—based on generating the same efficient frontier—that mean-CVaR optimization is equivalent to minimizing CVaR with a minimum expected profit and also to maximizing the expected profit subject to a CVaR constraint. Therefore, we decompose the firm's optimization into CVaR minimization via financial hedging with a minimum expected profit and profit maximization via operational flexibility subject to a CVaR constraint. This setup allows one to implement mean-CVaR optimization within a centralized functional structure, via (P1), or within a decentralized functional structure via (P2) and (P3). Mean-variance and expected utility models both typically require MNC to be a *centralized* planner of operations and finance using a single objective function. Such centralized models are accurate provided one accepts the rather strong assumption that incentives between operational and financial departments are perfectly aligned. In practice, most MNCs have a *decentralized* functional structure. Hence, one advantage of mean-CVaR optimization is that it allows us to decompose operations and finance: operational strategy focuses on maximizing expected profit subject to a CVaR constraint, while financial hedging focuses on minimizing CVaR subject to a minimum expected profit.

5.2.2 Optimal Financial Hedging Strategy

We first use the decomposed mean-CVaR optimization to optimize financial hedging decisions. Since the objective of maximizing mean-CVaR is equivalent to minimizing CVaR subject to a minimum expected profit, per Lemma 5.1, it follows that the portfolio of financial hedging contracts with respect to (w.r.t.) $\boldsymbol{h} \in \mathbb{R}^n$ for each $\boldsymbol{p} = (\boldsymbol{Q}, \boldsymbol{h}) \in \mathbf{P}$ can be optimized by solving the following problem:

$$\min_{\boldsymbol{h} \in \mathbb{R}^n} \mathrm{CVaR}_\beta(\boldsymbol{Q}, \boldsymbol{h}, \boldsymbol{y}) \tag{5.3}$$

$$\text{s.t.} \quad \mathbb{E}[\Pi(\boldsymbol{Q}, \boldsymbol{h}, \boldsymbol{y})] \geq \rho, \mathbb{E}[H(\boldsymbol{h})] = 0, \ \boldsymbol{p} \in \mathbf{P}$$

Here the no-arbitrage constraint $\mathbb{E}[H(\boldsymbol{h})] = 0$ indicates that the expected profit remains unaffected by financial hedging for a fixed capacity portfolio $\boldsymbol{Q}$. That is, financial hedging has no *direct* influence on expected profit; it can affect expected profit only if it alters the feasible set of capacity portfolios by minimizing CVaR (see Proposition 5.3 in Sect. 5.3).

In financial hedging only (i.e., in the absence of operational flexibility), the loss function is $l(\boldsymbol{Q}, \boldsymbol{h}, \boldsymbol{y}) = -\Pi^{\mathrm{fh}}(\boldsymbol{Q}, \boldsymbol{h}, \boldsymbol{y}) = -\Pi^{\mathrm{bc}} - H(\boldsymbol{h}) = -\sum_{i \in \{1,2\}} s_i(p_i - c_i)\min(D_i, K_i) - R(\boldsymbol{h}) + \left((c_1^k, sc_2^k, c_r^k)\boldsymbol{Q} + C(\boldsymbol{h})\right)e^{\gamma t}$, where $s_i = 1$ if $i = 1$ or $s_i = s$ if $i = 2$ and $\boldsymbol{Q} = (Q_1, Q_2, 0)^{\mathrm{T}}$. This loss function is *continuous* and linear w.r.t. $\boldsymbol{h}$. Hence

CVaR can be calculated by minimizing the following auxiliary function w.r.t. the variable $\alpha \in \mathbb{R}$ (Rockafellar and Uryasev 2000, 2002):

$$F_\beta(\boldsymbol{Q},\boldsymbol{h},\alpha) \triangleq \alpha + \frac{1}{1-\beta} \int_{\boldsymbol{y}\in\mathbb{R}^m} [l(\boldsymbol{Q},\boldsymbol{h},\boldsymbol{y}) - \alpha]^+ f(\boldsymbol{y})\mathrm{d}\boldsymbol{y}. \quad (5.4)$$

In this case, the global firm has neither reliable domestic capacity nor production switching, which leads to the linearity and continuity of the profit function (and loss function). Hence the optimal financial hedging portfolio here consists of forward contracts.

For integrated risk management (i.e., financial hedging with operational flexibility), real options lead to multiple *discrete* loss functions (see Proposition 5.2 in Sect. 5.3) whose corresponding probabilities arise from the support of continuous uncertainty domains (see Table 5.2, also in that section). In this case, operational flexibility can enlarge the feasible decision vector set w.r.t. the minimum expected profit. Here $\mathrm{CVaR}_\beta(\boldsymbol{Q},\boldsymbol{h},\boldsymbol{y})$ can be calculated as a weighted sum of the CVaR in each uncertainty domain and then optimized by sampling the probability density distribution $f(\boldsymbol{y})$; that distribution has support $\Omega_i (i = 1, 2, \ldots, 10)$, and the respective probabilities P_i generate a collection of vectors $\boldsymbol{y}_i$. Hence CVaR can be minimized via the formula

$$G_\beta(\boldsymbol{Q},\boldsymbol{h},\alpha) \triangleq \alpha + \frac{1}{1-\beta} \sum_{i=0}^{n} P_i [l(\boldsymbol{Q},\boldsymbol{h},\boldsymbol{y}_i) - \alpha]^+ \quad (5.5)$$

(cf. Rockafellar and Uryasev 2002), where n denotes the number of scenarios. So when the loss distribution is fat tailed at rare and extremely low exchange rates, the optimal financial hedging portfolio consists of deep out-of-the-money put options.

If financial markets are efficient and arbitrage-free, then hedging contracts are fairly priced; thus it is not possible to outperform the market systematically by earning a positive expected return (cf. Chowdhry and Howe 1999; Hommel 2003). In other words, financial hedging contracts are subject to the no-arbitrage constraint $(\mathbb{E}[H(\boldsymbol{h})] = 0)$ and so

$$\begin{aligned} \mathbb{E}[C(\boldsymbol{h})] &= \mathbb{E}[R(\boldsymbol{h})e^{-\gamma t}] \\ &= \begin{cases} \int_{\underline{s}}^{\bar{s}} h_f^{\mathrm{T}} (s_f - s) e(s) ds e^{-\gamma t} & \mathit{for} \quad l(\boldsymbol{Q},\boldsymbol{h},\boldsymbol{y}) = -\Pi^{\mathrm{fh}}(\boldsymbol{Q},\boldsymbol{h},\boldsymbol{y}), \\ \int_{\underline{s}}^{\bar{s}} \boldsymbol{h}_p^{\mathrm{T}} [\boldsymbol{s}_p - \boldsymbol{s}]^+ e(s) ds e^{-\gamma t} & \mathit{for} \quad l(\boldsymbol{Q},\boldsymbol{h},\boldsymbol{y}) = -\Pi^{\mathrm{int}}(\boldsymbol{Q},\boldsymbol{h},\boldsymbol{y}); \end{cases} \end{aligned} \quad (5.6)$$

here $[t]^+ = \max\{t, 0\}$, s_f is the contractual rate of currency forwards, and $\boldsymbol{s_p}$ represents the vector of strike prices in put options. In case of integrated risk management, operational flexibility results in multiple and discrete loss functions; hence optimal options contracts can have multiple strike prices. The global firm thus

Table 5.2 Optimal production quantities in Stage 2

Domain	Optimal Strategy	q_1^*	q_r^*	q_2^*
$\Omega_1(Q) = \{\mathbf{R} \in \mathbb{R}^2 \mid \sum K_i < \sum D_i, s \in [\underline{s}, \bar{s}]\}$	Full capacity sourcing with switching option	K_1	K_r	K_2
$\Omega_2(Q) = \{\mathbf{R} \in \mathbb{R}^2 \mid \sum D_i \leq K_2, s \in \left[\underline{s}, \frac{c_1 - c_{21}^s}{c_2}\right]\}$	Single sourcing from resource 2	0	0	$\sum D_i$
$\Omega_3(Q) = \{\mathbf{R} \in \mathbb{R}^2 \mid K_2 < \sum D_i \leq K_1 + K_2, s \in \left[\underline{s}, \frac{c_1 - c_{21}^s}{c_2}\right]\}$	Sourcing from resource 2 with hedge from 1	$\sum D_i - K_2$	0	K_2
$\Omega_4(Q) = \{\mathbf{R} \in \mathbb{R}^2 \mid \sum D_i \leq K_1 + K_2, s \in (\frac{c_1 - c_{21}^s}{c_2}, \frac{c_r - c_{21}^s}{c_2}]\}$	Splitting with priority hedge from 2 and switching option	$\min(D_1, K_1) + (D_2 - K_2)^+$	0	$\min(D_2, K_2) + (D_1 - K_1)^+$
$\Omega_5(Q) = \{\mathbf{R} \in \mathbb{R}^2 \mid \sum D_i \leq K_1 + K_2, s \in \left(\frac{c_r - c_{21}^s}{c_2}, \frac{c_1}{c_2 - c_{12}^s}\right]\}$	Splitting with priority hedge from reliable capacity and switching option	$\min(D_1, K_1) + (D_2 - K_2)^+$	$(D_1 - K_1)^+$	$\min(D_2, K_2) + (D_1 - K_1 - K_b)^+$
$\Omega_6(Q) = \{\mathbf{R} \in \mathbb{R}^2 \mid \sum D_i \leq K_1, s \in (\frac{c_1}{c_2 - c_{12}^s}, \bar{s}]\}$	Single sourcing from resource 1	$\sum D_i$	0	0
$\Omega_7(Q) = \{\mathbf{R} \in \mathbb{R}^2 \mid K_1 < \sum D_i \leq K_1 + K_2, s \in (\frac{c_1}{c_2 - c_{12}^s}, \frac{c_r}{c_2 - c_{12}^s}]\}$	Sourcing from resource 1 with hedge from 2	K_1	0	$\sum D_i - K_1$
$\Omega_8(Q) = \{\mathbf{R} \in \mathbb{R}^2 \mid K_1 < \sum D_i \leq K_1 + K_r, s \in (\frac{c_r}{c_2 - c_{12}^s}, \bar{s}]\}$	Sourcing from resource 1 with hedge from reliable capacity	K_1	$(\sum D_i - K_1)^+$	0
$\Omega_9(Q) = \{\mathbf{R} \in \mathbb{R}^2 \mid K_1 + K_r < \sum D_i \leq \sum K_i, s \in \left(\frac{c_r}{c_2 - c_{12}^s}, \bar{s}\right)\}$	Sourcing from resource 1 and reliable capacity with hedge from 2	K_l	K_r	$\sum D_i - K_l - K_r$
$\Omega_{10}(Q) = \{\mathbf{R} \in \mathbb{R}^2 \mid K_1 + K_2 < \sum D_i \leq \sum K_i, s \in \left[\underline{s}, \frac{c_r + c_{21}^s}{c_2}\right]\}$	Sourcing from resources 1 and 2 with hedge from reliable capacity	K_1	$\sum D_i - K_1 - K_2$	K_2

tailors its CVaR through a hedge portfolio of forwards or put options, as detailed in the analysis to follow. We can therefore write the magnitude of the monetary flow transferred via financial hedging as

$$\int_{\underline{s}}^{s_i} (R(\boldsymbol{h})e^{-\gamma t} - C(\boldsymbol{h}))e(s)\mathrm{d}s = \int_{s_i}^{\bar{s}} (R(\boldsymbol{h})e^{-\gamma t} - C(\boldsymbol{h}))e(s)\mathrm{d}s, \tag{5.7}$$

where $i = f$ for $l(\boldsymbol{Q}, \boldsymbol{h}, \boldsymbol{y}) = -\Pi^{\text{fh}}(\boldsymbol{Q}, \boldsymbol{h}, \boldsymbol{y})$ and $i = p$ for $l(\boldsymbol{Q}, \boldsymbol{h}, \boldsymbol{y}) = -\Pi^{\text{int}}(\boldsymbol{Q}, \boldsymbol{h}, \boldsymbol{y})$. That is, the firm hedges financially through forwards (resp., put options) in the case of financial hedging only (resp., integrated risk management). The left-hand side of Eq. (5.7) represents the expected positive payoff from financial hedging when there is a shortfall in operating cash flow; the right-hand side (RHS) represents the expected negative payoff (i.e., the costs and losses) of financial hedging when there is no shortfall in operating profit. Equation (5.7) indicates that financial hedging optimizes CVaR by transferring monetary flows—across the probability space—from voluntary to binding positions.

Proposition 5.1 *The optimal financial hedging strategy can be derived as follows:*

$$\begin{aligned} \boldsymbol{h}^* &= \arg\min_{\boldsymbol{h}\in\mathbb{R}^n} \mathrm{CVaR}_\beta(\boldsymbol{Q}, \boldsymbol{h}, \boldsymbol{y}) \\ &= \begin{cases} \arg\min_{\boldsymbol{h}\in\mathbb{R}^n} F_\beta(\boldsymbol{Q}, \boldsymbol{h}, \alpha) & \textit{for} \quad l(\boldsymbol{Q}, \boldsymbol{h}, \boldsymbol{y}) = -\Pi^{\text{fh}}(\boldsymbol{Q}, \boldsymbol{h}, \boldsymbol{y}) \\ \arg\min_{\boldsymbol{h}\in\mathbb{R}^n} G_\beta(\boldsymbol{Q}, \boldsymbol{h}, \alpha) & \textit{for} \quad l(\boldsymbol{Q}, \boldsymbol{h}, \boldsymbol{y}) = -\Pi^{\text{int}}(\boldsymbol{Q}, \boldsymbol{h}, \boldsymbol{y}) \end{cases}, \end{aligned} \tag{5.8}$$

$$\text{s.t.} \quad \mathbb{E}[\Pi(\boldsymbol{Q}, \boldsymbol{h}, \boldsymbol{y})] \geq \rho, \mathbb{E}[H(\boldsymbol{h})] = 0.$$

Moreover, if the optimal solution to financial hedging is not (necessarily) unique, then

$$(\boldsymbol{h}^*, \mathbb{E}^*[\Pi(\boldsymbol{Q}, \boldsymbol{h}, \boldsymbol{y})]) \in \arg\min_{\boldsymbol{h}\in\mathbb{R}^n} \mathrm{CVaR}_\beta(\boldsymbol{Q}, \boldsymbol{h}, \boldsymbol{y}) \Leftrightarrow$$

$$\boldsymbol{h}^* \in \arg\min_{\boldsymbol{h}\in\mathbb{R}^n} \mathrm{CVaR}_\beta(\boldsymbol{Q}, \boldsymbol{h}, \boldsymbol{y}), \mathbb{E}^*[\Pi(\boldsymbol{Q}, \boldsymbol{h}, \boldsymbol{y})] \in \arg\max_{\boldsymbol{h}\in\mathbb{R}^n} \mathbb{E}[\Pi(\boldsymbol{Q}, \boldsymbol{h}^*, \boldsymbol{y})].$$

According to Proposition 5.1, the MNC should adopt financial hedging whenever its operating cash flow falls below the VaR threshold that minimizes profit shortfall (i.e., CVaR). If multiple solutions to CVaR minimization are evident, then the decision vector associated with the highest expected profit will be chosen. The payoff from financial hedging aims to replicate (offset), based on a conditional expectation, the size of CVaR in the corresponding exchange rate interval.

5.2.3 *Optimal Operational Flexibility Strategy*

We proceed in this section to optimize the global firm's operational flexibility strategy. By Lemma 5.1, optimizing mean-CVaR is equivalent to maximizing expected profit subject to a CVaR constraint; hence the firm's operational decisions w.r.t. $\boldsymbol{Q} \in \mathbb{R}^3_+$ for $\boldsymbol{p} = (\boldsymbol{Q}, \boldsymbol{h}) \in \mathbf{P}$ can be optimized by solving

$$\max_{\boldsymbol{Q} \in \mathbb{R}^3_+} \mathbb{E}[\Pi(\boldsymbol{Q}, \boldsymbol{h}, \boldsymbol{y})] \tag{5.9}$$

$$\text{s.t.} \quad \mathrm{CVaR}_\beta(\boldsymbol{Q}, \boldsymbol{h}, \boldsymbol{y}) \leq \omega, \boldsymbol{p} \in \mathbf{P}.$$

We now use backward induction to characterize the global firm's optimal production planning and capacity investment in, respectively, Stages 2 and 1.

Stage 2: Production planning

Proposition 5.2 *The optimal production quantities $\boldsymbol{q}_i^*$ in Stage 2 have one of the ten distinct forms listed in Table 5.2, where the applicable form depends on the exchange rate and the realized capacity–demand matching state.*

The cash flow in Stage 2 resulting from operational decisions is

$$\Pi^{\mathrm{int}}(\boldsymbol{Q}, \boldsymbol{y}) = (p_1, sp_2)\boldsymbol{d}_j^{*\mathrm{T}} - (c_1, sc_2, c_r)\boldsymbol{q}_i^{*\mathrm{T}} - \left(c_{21}^s, sc_{12}^s\right)\boldsymbol{q}_{ij}^{s*\mathrm{T}} - \left(c_1^k, sc_2^k, c_r^k\right)\boldsymbol{Q}e^{\gamma t}. \tag{5.10}$$

Table 5.2 summarizes the optimal operational flexibility strategies in various domains with support Ω_i $(i = 1, 2, \ldots, 10)$. In response to specific realizations of uncertainty, the global firm can adjust the production quantities derived from three different resources and use switching options to leverage production cost variations resulting from fluctuations in the exchange rate.

Stage 1: Capacity investment

The optimal solutions to capacity investment can be derived by trading off capacity costs against weighted capacity shadow prices $\mu(Q, \Omega_i)$ (i.e., marginal profit from adding unit capacity). Using necessary and sufficient Kuhn–Tucker conditions, we establish the following result.

Proposition 5.3 *A capacity vector $\boldsymbol{Q}^* \in \mathbb{R}^3_+$ is optimal if and only if there exists a $\boldsymbol{\delta} \in \mathbb{R}^3_+$ such that*

$$c_i^k e^{\gamma t} - \boldsymbol{\delta} = \sum P(\Omega_i(\boldsymbol{Q}^*))\boldsymbol{\mu}(\boldsymbol{Q}, \Omega_i) \text{ and } \boldsymbol{\delta}^{\mathrm{T}}\boldsymbol{Q}^* = 0, \tag{5.11}$$

$$s.t.\ \mathrm{CVaR}_\beta(\boldsymbol{Q}, \boldsymbol{h}, \boldsymbol{y}) \le \omega.$$

Proposition 5.3 indicates that the global firm's optimal amount of capacity investment is such that the marginal benefits of the resulting capacities—weighted by the probability of each scenario—match the marginal costs of those capacities.

5.3 Interplay Between Operations and Finance

Next, we explore the relationships among capacity reshoring, production switching, and financial hedging when there are mean-CVaR trade-offs.

5.3.1 Relationship Between Capacity Reshoring and Production Switching

Reliable domestic production hedges supply risk, whereas production switching hedges both exchange rate and supply risks. Hence the natural question arises: Are the two operational strategies complements or substitutes? In order to assess their relationship, we first examine the joint and respective values of production switching and capacity reshoring.

Definition 5.2 The *value* of operational flexibility refers to the utility increment—w.r.t. the mean-CVaR objective and relative to the base case—from adopting the strategy of capacity reshoring and/or production switching.

Proposition 5.4 *The value of reliable production only is*

$$\Delta U[\Pi^{\mathrm{rp}}] = (1+\lambda)\sum P(\Omega_i(\boldsymbol{Q}^*))(q_r^*(p_1 - c_r) - c_r^k K_r). \tag{5.12}$$

The value of switching options only is

$$\Delta U[\Pi^{\mathrm{sw}}] = (1+\lambda)\sum P(\Omega_i(\boldsymbol{Q}^*))v_{ij}^*(s_i c_i - s_j c_j - s_j c_{ij}^s); \tag{5.13}$$

here $i \neq j$, $s_i = \begin{cases} 1, i = 1 \\ s, i = 2 \end{cases}$, and $s_j = \begin{cases} 1, j = 1 \\ s, j = 2 \end{cases}$.

The value of full operational flexibility is

$$\Delta U[\Pi^{\mathrm{op}}] = (1+\lambda)(\mathbb{E}[\Pi^{\mathrm{op}}] - \mathbb{E}[\Pi^{\mathrm{bc}}]). \tag{5.14}$$

Definition 5.3 Capacity reshoring and production switching are complements, separate, or substitutes according as whether (respectively)

$\Delta U[\Pi^{\mathrm{op}}] > \Delta U[\Pi^{\mathrm{sw}}] + \Delta U[\Pi^{\mathrm{rp}}]$, $\Delta U[\Pi^{\mathrm{op}}] = \Delta U[\Pi^{\mathrm{sw}}] + \Delta U[\Pi^{\mathrm{rp}}]$, or $\Delta U[\Pi^{\mathrm{op}}] < \Delta U[\Pi^{\mathrm{sw}}] + \Delta U[\Pi^{\mathrm{rp}}]$.

Both capacity reshoring and production switching represent excess capacity that hedges mismatch risk and exchange rate uncertainty; it is therefore intuitive that they are substitutes. However, the following theorem shows that they can also be complements under certain conditions.

Theorem 5.1 *Capacity reshoring and production switching are complements (resp., substitutes) when they are used to fulfill foreign (resp., domestic) demand. The interaction effects between reliable domestic production and switching options (i.e., the value increment in utility function) are expressed formally as*

$$\begin{aligned}
\Delta U[\mathrm{sw},\mathrm{rp}] &\equiv \Delta U[\Pi^{\mathrm{op}}] - \Delta U[\Pi^{\mathrm{sw}}] - \Delta U[\Pi^{\mathrm{rp}}] = P\left(\frac{p_1 + c_{21}^s}{p_2} < s \le \bar{s}\right)\min\left(q_r^*, v_{21}^*\right)\left|sc_2 - c_r - c_{21}^s\right| \\
&+ P\left(\frac{c_r}{c_2 + c_{12}^s} \le s < \frac{p_1}{c_{12}^s + p_2}\right)\left\{\min\left([D_1 - K_1 - K_r]^+, K_2\right) - \min\left([D_1 - K_1]^+, K_2\right)\right\}\left(p_1 - sp_2 - sc_{12}^s\right) \\
&+ P\left(s_t \le s < \frac{c_r}{c_2 + c_{12}^s}\right)\left\{\min\left([D_1 - K_1 - K_2]^+, K_r\right) - \min\left([D_1 - K_1]^+, K_r\right)\right\}\left(p_1 - c_r\right) \\
&+ (1+\lambda)P(\underline{s} \le s < s_t)\left\{\min\left([D_1 - K_1 - K_2]^+, K_r\right) - \min\left([D_1 - K_1]^+, K_r\right)\right\}\left(p_1 - c_r\right)
\end{aligned} \tag{5.15}$$

Here $s_t = \max\{s \in [\underline{s}, \bar{s}] : l(\boldsymbol{Q}, \boldsymbol{h}, \boldsymbol{y}) = -\Pi^{\mathrm{int}}(\boldsymbol{Q}, \boldsymbol{h}, \boldsymbol{y}) < \alpha\}$. The first term on the RHS of (5.15) is nonnegative and represents the complementary effects when switching options use reliable domestic capacity to fulfill *foreign* demand D_2. In contrast, the second, third, and fourth terms on the RHS of (5.15) are nonpositive; they capture the substitution effects when both reliable capacity and production switching are used to fulfill *domestic* demand D_1. Thus the relationship between capacity reshoring and production switching depends on whether the complementary or substitution effects are more significant. Theorem 5.1 indicates that higher foreign (domestic) demand expectation encourages (discourages) joint adoption of capacity reshoring and production switching. Hence the relative magnitudes of domestic and foreign markets determine the optimal portfolio of operational flexibility.

5.3.2 *Relationship Between Operational Flexibility and Financial Hedging*

We now assess the interaction between operations and finance by examining the value of financial hedging in both the presence and the absence of operational flexibility. The *value* of financial hedging in the absence (resp., presence) of operational flexibility is the increment in the mean-CVaR utility relative to the base case (resp., to full operational flexibility) that is created via currency derivatives. Recall that (no-arbitrage) financial hedging yields $\mathbb{E}[\mathrm{H}(\boldsymbol{h})] = 0$, from which it follows that the value of optimal financial hedging w.r.t. the base case can be readily derived as $\Delta U[\mathrm{FH}^{\mathrm{bc}}] \equiv U[\Pi^{\mathrm{fh}}] - U[\Pi^{\mathrm{bc}}] = \lambda(\mathrm{CVaR}_\beta^{\mathrm{bc}} - \mathrm{CVaR}_\beta^{\mathrm{fh}}) + (\mathbb{E}[\Pi^{\mathrm{fh}}] - \mathbb{E}[\Pi^{\mathrm{bc}}])$. The value of optimal financial hedging under integrated risk management (relative to full operational flexibility) can likewise be written as $\Delta U[\mathrm{FH}^{\mathrm{op}}] \equiv U[\Pi^{\mathrm{int}}] - U[\Pi^{\mathrm{op}}] = \lambda(\mathrm{CVaR}_\beta^{\mathrm{op}} - \mathrm{CVaR}_\beta^{\mathrm{int}}) + (\mathbb{E}[\Pi^{\mathrm{int}}] - \mathbb{E}[\Pi^{\mathrm{op}}])$. These expressions show that the value of financial hedging with or without operational flexibility consists of two factors: risk reduction effects and profit enhancement effects.

Definition 5.4 Operational flexibility and financial hedging are complements, separate, or substitutes according as whether (respectively) $\Delta U[\mathrm{FH}^{\mathrm{op}}] > \Delta U[\mathrm{FH}^{\mathrm{bc}}]$, $\Delta U[\mathrm{FH}^{\mathrm{op}}] = \Delta U[\mathrm{FH}^{\mathrm{bc}}]$, or $\Delta U[\mathrm{FH}^{\mathrm{op}}] < \Delta U[\mathrm{FH}^{\mathrm{bc}}]$.

Theorem 5.2 *Operational flexibility and financial hedging are substitutes in risk reduction. These two mechanisms can be complements in profit enhancement when financial hedging increases the feasible set of capacity portfolios by relaxing the CVaR constraint. Their interaction effects (i.e., the value increment in the firm's utility) are given as follows:*

$$\begin{aligned}\Delta U[\mathrm{fh},\mathrm{op}] = {}& \lambda\left(\mathrm{CVaR}_\beta^{\mathrm{op}} - \mathrm{CVaR}_\beta^{\mathrm{bc}} + \mathrm{CVaR}_\beta^{\mathrm{fh}} - \mathrm{CVaR}_\beta^{\mathrm{int}}\right) \\ & + (\mathbb{E}[\Pi^{\mathrm{bc}}] - \mathbb{E}[\Pi^{\mathrm{fh}}] + \mathbb{E}[\Pi^{\mathrm{int}}] - \mathbb{E}[\Pi^{\mathrm{op}}]).\end{aligned} \tag{5.16}$$

This theorem states that the interaction effects between operational flexibility and financial hedging are determined by two factors, the first of which is the risk reduction interaction: $\lambda\left(\mathrm{CVaR}_\beta^{\mathrm{op}} - \mathrm{CVaR}_\beta^{\mathrm{bc}} + \mathrm{CVaR}_\beta^{\mathrm{fh}} - \mathrm{CVaR}_\beta^{\mathrm{int}}\right) < 0$. Therefore, operational flexibility and financial hedging are substitutes w.r.t. risk reduction. The second factor is the profit enhancement interaction: $\mathbb{E}[\Pi^{\mathrm{bc}}] - \mathbb{E}[\Pi^{\mathrm{fh}}] + \mathbb{E}[\Pi^{\mathrm{int}}] - \mathbb{E}[\Pi^{\mathrm{op}}] > 0$ if and only if $\mathbb{E}[\Pi^{\mathrm{int}}] - \mathbb{E}[\Pi^{\mathrm{op}}] > E[\Pi^{\mathrm{fh}}] - \mathbb{E}[\Pi^{\mathrm{bc}}]$. It follows that operational flexibility and financial hedging can be complements w.r.t. profit enhancement.

5.3.3 *Efficient Frontier Analysis*

In order to demonstrate the interactions of operational flexibility and financial hedging, we conduct an efficient frontier analysis following the approach of van Mieghem (2003, p. 295). Our primary motivation for adopting this approach is that we used the equivalence of efficient frontiers in Lemma 5.1, i.e., optimizing mean-CVaR is equivalent to minimizing CVaR with a minimum expected profit and also to maximizing the expected profit under a CVaR constraint.

Definition 5.5 A portfolio of real and financial assets $\boldsymbol{p} = (\boldsymbol{Q}, \boldsymbol{h})$ is *efficient* if and only if there does not exist another portfolio $\boldsymbol{p}'$ such that $U[\Pi(\boldsymbol{p}')] > U[\Pi(\boldsymbol{p})]$.

Definition 5.6 The efficient frontier $\mathcal{F}^{x}_{CVaR}$ for strategy $x = \text{bc}, \text{rp}, \text{sw}, \text{op}, \text{fh}, \text{int}$ (see Definition 5.1) is the set of mean-CVaR pairs of efficient portfolios:

$$\mathcal{F}^{x}_{CVaR} = \{(\mathbb{E}[\Pi^{x}(\boldsymbol{p})], \text{CVaR}^{x}_{\beta}(\boldsymbol{p}))|\boldsymbol{p} \text{ is efficient}\}.$$

The global firm's initial frontier of its mean-CVaR objective is denoted by $\mathcal{F}^{\text{bc}}_{CVaR}$ in Fig. 5.3. Adopting operational flexibility and/or financial hedging enables the firm to improve its utility by deploying real and financial assets to shift northwest the efficient (Pareto-optimal) frontier of its portfolio $\boldsymbol{p}$. We use $\mathbb{E}[\Pi(X)]$ for $X = \text{B}, \text{E}, \text{F}$ to denote the expected profit when maximum utility is attained at points B, E, and F; here $\mathbb{E}[\Pi(\text{B})] > E[\Pi(\text{C})] > E[\Pi(\text{A})] > E[\Pi(\text{D})] > \rho$. Observe that $\text{CVaR}_{\beta}(\text{G})$ and $\text{CVaR}_{\beta}(X)$for $X = \text{E}, \text{F}$ signify the minima of CVaR attainable via, respectively, financial hedging only (at point G) and integrated risk management (at points E or F).

In the base case, even though the firm's utility can be maximized at point A, the risk-averse firm would choose point D by virtue of the CVaR constraint because $\text{CVaR}_{\beta}(\text{D}) = \omega^{\text{bc}}$ and $\mathbb{E}[\Pi(\text{D})] > \rho$. In the case of financial hedging only, the efficient frontier is shifted *due* west to $\mathcal{F}^{\text{fh}}_{CVaR}$ by vector $\overrightarrow{\text{AG}}$, where $\text{CVaR}_{\beta}(\text{G}) < \omega$; in this case, the utility-maximizing point G can be selected. Hence the value of

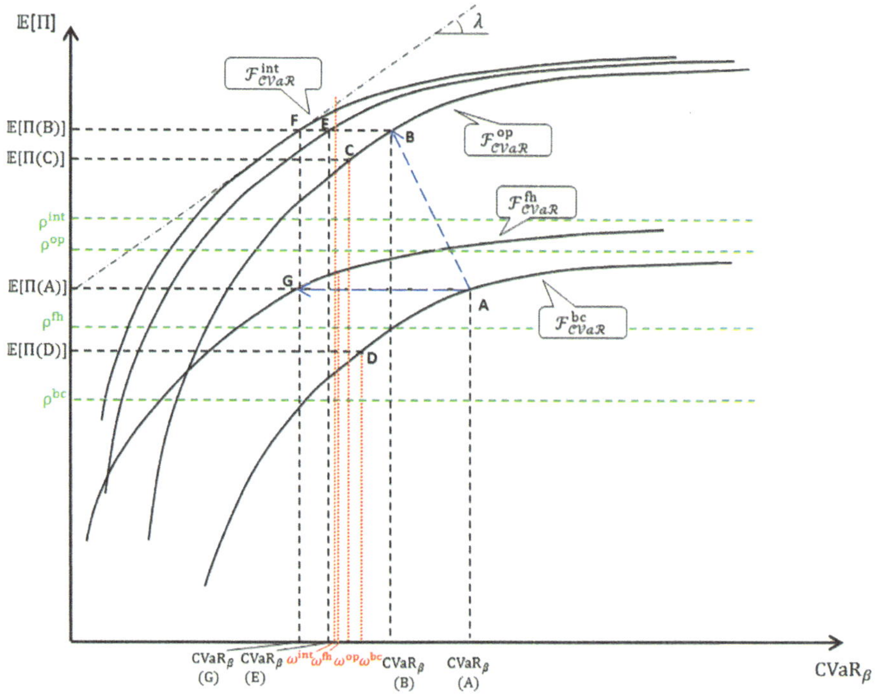

Fig. 5.3 Efficient frontier analysis: Interaction between operations and finance

optimal financial hedging without operational flexibility is $\Delta U\left[\mathrm{FH}^{\mathrm{bc}}\right] = \lambda\left(\mathrm{CVaR}_{\beta}^{\mathrm{bc}} - \mathrm{CVaR}_{\beta}^{\mathrm{fh}}\right) + \mathbb{E}\left[\Pi^{\mathrm{fh}}\right] - \mathbb{E}\left[\Pi^{\mathrm{bc}}\right]$. Under full operational flexibility, real options shift the efficient frontier northwest to $\mathcal{F}_{\mathcal{CVaR}}^{\mathrm{op}}$ by vector $\overrightarrow{\mathrm{AB}}$, where the firm's utility can be maximized at point B. However, the risk-averse firm would choose point C as the optimal feasible point when considering the constraint on CVaR, where $\mathrm{CVaR}_{\beta}(\mathrm{C}) = \omega^{\mathrm{op}}$.

In integrated risk management, we distinguish two cases. First, if financial hedging reduces CVaR to point E, where $\mathrm{CVaR}_{\beta}(\mathrm{E}) = \mathrm{CVaR}_{\beta}^{\mathrm{int}} > \mathrm{CVaR}_{\beta}(\mathrm{G}) = \mathrm{CVaR}_{\beta}^{\mathrm{fh}}$, then operational flexibility and financial hedging are substitutes in risk reduction because of their interaction effects w.r.t. CVaR are $\lambda(\mathrm{CVaR}_{\beta}^{\mathrm{op}} - \mathrm{CVaR}_{\beta}^{\mathrm{bc}} + \mathrm{CVaR}_{\beta}^{\mathrm{fh}} - \mathrm{CVaR}_{\beta}^{\mathrm{int}}) < 0$. Second, if financial hedging reduces CVaR to point F, where $\mathrm{CVaR}_{\beta}(\mathrm{F}) = \mathrm{CVaR}_{\beta}^{\mathrm{int}} = \mathrm{CVaR}_{\beta}(\mathrm{G}) = \mathrm{CVaR}_{\beta}^{\mathrm{fh}}$, then operational flexibility and financial hedging are substitutes in risk reduction given that their interaction effects w.r.t. CVaR are $\lambda(\mathrm{CVaR}_{\beta}^{\mathrm{op}} - \mathrm{CVaR}_{\beta}^{\mathrm{bc}}) < 0$. Yet the profit enhancement interaction, $\mathbb{E}\left[\Pi^{\mathrm{bc}}\right] - \mathbb{E}\left[\Pi^{\mathrm{fh}}\right] + \mathbb{E}\left[\Pi^{\mathrm{int}}\right] - \mathbb{E}[\Pi^{\mathrm{op}}]$, depends on the expected profit increments from financial hedging with and without operational flexibility as follows. If $\mathbb{E}\left[\Pi^{\mathrm{int}}\right] - \mathbb{E}[\Pi^{\mathrm{op}}] > E\left[\Pi^{\mathrm{fh}}\right] - \mathbb{E}\left[\Pi^{\mathrm{bc}}\right]$, then operational flexibility and financial hedging are complements in profit enhancement; if $\mathbb{E}\left[\Pi^{\mathrm{int}}\right] - \mathbb{E}[\Pi^{\mathrm{op}}] \leq \mathbb{E}\left[\Pi^{\mathrm{fh}}\right] - \mathbb{E}\left[\Pi^{\mathrm{bc}}\right]$, then operational flexibility and financial hedging are strictly substitutes owing to their interaction effects (Theorem 5.2): $\Delta U[\mathrm{fh}, \mathrm{op}] < 0$.

We derive three managerial insights from this efficient frontier analysis. First, operational flexibility and financial hedging can be complements. On the one hand, operational flexibility enhances expected profit and reduces downside risk; financial hedging *minimizes* downside risk yet can affect the feasible set of capacity portfolios—and thereby profitability—only by relaxing a CVaR constraint. This result is in line with the literature on integrated risk management (Smith and Stulz 1985; Mello et al. 1995, 1996). On the other hand, operational flexibility can be viewed as real options written on costs in foreign/domestic currency whereas financial hedging via put options on foreign currency mitigates the impact of rare and extreme low exchange rates on cash flow.

Second, operational flexibility and financial hedging can be substitutes. With operational flexibility, financial options hedge only rare and extreme exchange rates because greater profits can be realized by instead using real options otherwise. Financial hedging has greater risk reduction effects when used alone. This result is consistent with empirical evidence on the interaction between financial and operational hedging, which shows a significant reduction in the use of financial derivatives as firms adopt more real options to manage currency risk (Aabo and Simkins 2005; Kim et al. 2006).

Third, collaboration and coordination between operations and finance is needed to minimize substitution effects in mean-CVaR utility, especially in risk reduction. Hence a two-way information exchange between operations and finance is crucial:

efficient financial hedging is based on precise estimation of cash flow distribution as determined by operational flexibility, while the feasibility of capacity choices relies—by virtue of the CVaR constraint—on financial hedging.

5.4 Numerical Examples

This section illustrates our main results by way of an extensive numerical study. Examples are provided not only to demonstrate the relative effectiveness of operational flexibility and financial hedging as well as the effects of exchange rate volatility and demand volatility but also to compare mean-CVaR and mean-variance objectives.

The numerical study relies on optimization via simulation using Palisade @RISK software with 32,000 uncertainty scenarios (which vary by exchange rates and supply–demand matching states) and the following benchmark parameter values: $p_i = 10, c_i = 7, c_r = 7.5, c_{ij}^s = 1, c_i^k e^{\gamma t} = 0.6, c_r^k e^{\gamma t} = 1.2, \lambda = 0.05$, and $R_i \sim \text{Unif}(0.91, 1)$ for $i, j = 1, 2$ with $i \neq j$. Because the difference between VaR and CVaR as a risk measure is negligible for normal distributions (Sarykalin et al. 2008, p. 280), for the exchange rate we choose a uniform distribution, $s \sim \text{Unif}(0.55, 1.45)$ (cf. Chowdhry and Howe 1999), whose interval varies with its coefficient of variation. Demands D_1 and D_2 are each uniformly distributed, $D_i \sim \text{Unif}(25, 125)$ for $i = 1, 2$, where again the interval varies with the coefficient of variation. Supply, demand, and exchange rates are independent, and they jointly characterize the ten uncertainty domains defined in Table 5.3.

Table 5.3 verifies Theorem 5.2 by showing the interaction effects between operational flexibility and financial hedging. In this benchmark example, the coefficient of variation in demand is $\delta(D_i) = 0.49$ and the exchange rate's coefficient of variation is $\delta(s) = 0.22$. Integrated risk management improves the firm's utility more significantly: by 15.76%, as compared with the utility increment due to full operational flexibility (13.68%) or to financial hedging only (1.85%). The implication is that operational flexibility and financial hedging are complements because the value increment from integrated risk management (15.76%) exceeds the *sum* of value increments (15.53%) from full operational flexibility and financial hedging only. Operational flexibility increases expected profit by 14.24%, whereas financial hedging has only a minor effect (2.14%) owing to the no-arbitrage assumption; recall that financial hedging can enhance expected profit only if it enlarges the feasible set of capacity portfolios (i.e., by relaxing a CVaR constraint). As discussed in Sect. 4.2, operational flexibility provides the benefits of profit enhancement and risk reduction; financial hedging minimizes risk, and its ability to enhance profits depends on whether the CVaR constraint is relaxed. For a firm adopting a single strategy, the relative effectiveness of these two choices depends on the relative magnitudes of the resulting profit enhancement and risk reduction, which in turn are functions of switching costs and the degree of risk aversion. The

Table 5.3 Relative effectiveness of operational flexibility and financial hedging

Base case			**Full operational flexibility**			$(\mathbf{T}^{op} - \mathbf{T}^{bc})/\mathbf{T}^{bc}(\mathbf{T} = \mathbf{U}^*_{0.95}, \mathbb{E}^*, \mathbf{CVaR}^*_{0.95})$		
$U^*_{0.95}$	$\mathbb{E}^*$	$\mathrm{CVaR}^*_{0.95}$	$U^*_{0.95}$	$\mathbb{E}^*$	$\mathrm{CVaR}^*_{0.95}$	$U^*_{0.95}$	$\mathbb{E}^*$	$\mathrm{CVaR}^*_{0.95}$
200.33	205.17	−96.83	227.75	234.39	−132.93	13.68%	14.24%	37.29%
Financial hedging only			**Integrated risk management**			$(\mathbf{T}^{int} - \mathbf{T}^{fh})/\mathbf{T}^{fh}$		
204.03	209.57	−110.76	231.91	238.73	−136.49	13.66%	13.91%	23.22%
$(\mathbf{T}^{fh} - \mathbf{T}^{bc})/\mathbf{T}^{bc}$			$(\mathbf{T}^{int} - \mathbf{T}^{op})/\mathbf{T}^{op}$			$(\mathbf{T}^{int} - \mathbf{T}^{bc})/\mathbf{T}^{bc}$		
1.85%	2.14%	14.39%	1.83%	1.85%	2.67%	15.76%	16.36%	40.96%

risk reduction effect from integrated risk management (40.96%) is less than the sum (51.68%) of reduction effects from full operational flexibility (37.29%) and financial hedging only (14.39%), a result that is due to the latters' substitution effects. We therefore conclude that operational flexibility and financial hedging are substitutes in risk reduction.

5.4.1 Impact of Exchange Rate Variability

We next examine how exchange rate volatility affects the global firm's capacity investment and performance for the six strategies described in Definition 5.1. When adjusting the exchange rate's volatility, we allow its coefficient of variation $\delta(s)$ to vary between 0.17 and 0.26 while keeping the expectation $\mathbb{E}[s]$ at a constant level of 1.05 (in our one-period setting, exchange rate variability is assumed to move within a certain range—for example, 30%; cf. de Grauwe 1988). The effects of increasing exchange rate volatility are plotted in Figs. 5.4a–d.

Figure 5.4a shows that optimal total capacity is decreasing in exchange rate volatility. Absent financial hedging, a capacity hedge (via real options) is the only risk management instrument; this explains why, given the CVaR constraint, optimal capacity declines rapidly as exchange rate volatility increases. When financial hedging is employed—that is, under financial hedging only (blue curves in

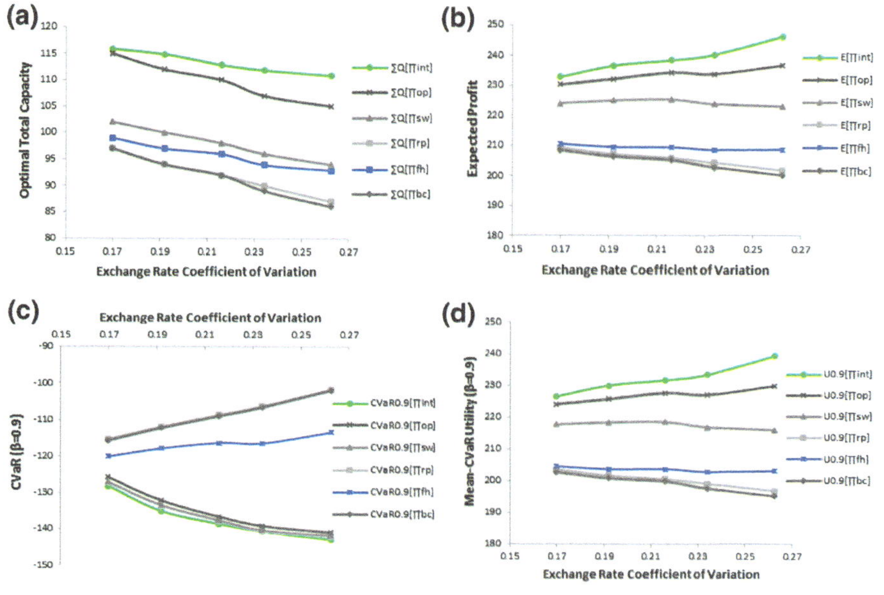

Fig. 5.4 Effect of exchange rate volatility on **a** optimal capacity; **b** expected profit; **c** CVaR; and **d** mean-CVaR

Fig. 5.4) or integrated risk management (green curves)—optimal total capacity is less sensitive to exchange rate volatility because the currency exposure in CVaR is significantly reduced by financial hedging. Figure 5.4b shows that the firm's expected profit is increasing in exchange rate volatility under full operational flexibility or integrated risk management, that the expected profit is much less affected under financial hedging only or switching options only, and that the expected profit is decreasing in exchange rate volatility for the base case and also for reliable production only. These findings underscore that more flexibility or more hedging (or both) increases the firm's ability to mitigate currency risk.

Figure 5.4c illustrates the risk reduction effects of financial hedging and real options in response to greater exchange rate volatility. The CVaR increases significantly with exchange rate volatility for the base case and for reliable production only, but that increase is much less under financial hedging only. In contrast, the CVaR is significantly *decreasing* in exchange rate volatility for firms that employ switching options only, full operational flexibility, or integrated risk management. These results indicate that production switching becomes more effective at reducing risk as exchange rate volatility increases. The expected utilities of firms that use production switching increase with exchange rate volatility but decrease (albeit at a slower rate) without production switching (i.e., for the base case, financial hedging only, and reliable production only); see Fig. 5.4d.

In order to compare the difference between mean-CVaR and mean-variance models, we conduct a numerical study by optimizing the MV objective ceteris paribus; see Fig. 5.5. The effects of exchange rate volatility on optimal total capacity, expected profit, and MV utility in panels (a), (b), and (d) of Fig. 5.5 are analogous to those under the mean-CVaR model. Figure 5.5c shows that variance is *nonmonotonic* in exchange rate volatility under financial hedging only or integrated risk management. This result differs from that in Fig. 5.4c, where CVaR is (w.r.t. exchange rate volatility) *decreasing* in integrated risk management but *increasing* in financial hedging only. This difference stems from the use of CVaR as a downside risk measure. Financial hedging via CVaR truncates below-threshold payoffs in a more conservative manner, whereas variance hedging tailors both upside potential and downside risk. We compare our results also with those from the two-capacity MV model of Ding et al. (2007), who find that variance is *slightly increasing* (w.r.t. exchange rate standard deviation) in financial hedging only or integrated risk management. The difference here lies in our incorporation of reliable domestic production as the third capacity. However, the result in Fig. 5.5c echoes the finding of Ding et al. (2007) in that financial hedging makes profit variance less sensitive to a reduction in optimal capacity.

5.4.2 Impact of Demand Variability

In order to study the effect of demand volatility, we vary the coefficient of variation in demand, $\delta(D_i)$, from 0.47 to 0.53 while keeping its expectation, $\mathbb{E}[D_i]$, at a

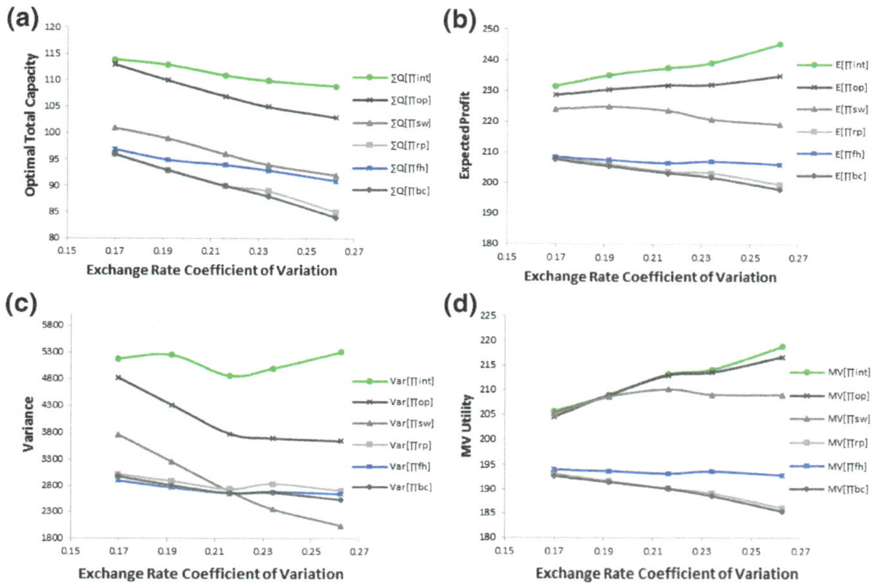

Fig. 5.5 Effect of exchange rate volatility on **a** optimal capacity; **b** expected profit; **c** variance; and **d** MV utility

constant level of 41.8. Figure 5.6a shows that optimal total capacity is decreasing in demand volatility whereas financial hedging makes capacity investment (slightly) less sensitive to demand volatility. The latter result follows because financial hedging reduces currency risk and thereby alters the feasible set of capacity portfolios w.r.t. the CVaR constraint; hence demand volatility will have less of an impact on optimal total capacity. In Fig. 5.6b we can see that expected profit declines owing to the reduced optimal capacity under all six strategies—an effect that differs from the one shown in Fig. 5.4b.

In Fig. 5.6c, CVaR increases with demand volatility under all six strategies—contra the results in Fig. 5.4c. This increase is slightly less significant with than without financial hedging, which is consistent with demand uncertainty being a *private* risk that cannot be hedged in financial markets (as argued by Chen et al. 2007b). Figure 5.6c reveals that, in comparison with Fig. 5.4c, financial hedging and production switching are less effective in reducing demand volatility risk than in reducing exchange rate risk. Finally, Fig. 5.6d illustrates that the net effect on mean-CVaR is decreasing in demand volatility. The presence of financial hedging slows down the decline in mean-CVaR utility because optimal total capacity becomes relatively insensitive to demand volatility.

For the purpose of mean-CVaR comparisons, Fig. 5.7 shows results from the mean-variance model (ceteris paribus). The effects of demand volatility on optimal total capacity, expected profit, and MV utility—as plotted in panels (a), (b), and (d) of the figure—are analogous to those under the mean-CVaR model. Figure 5.7c

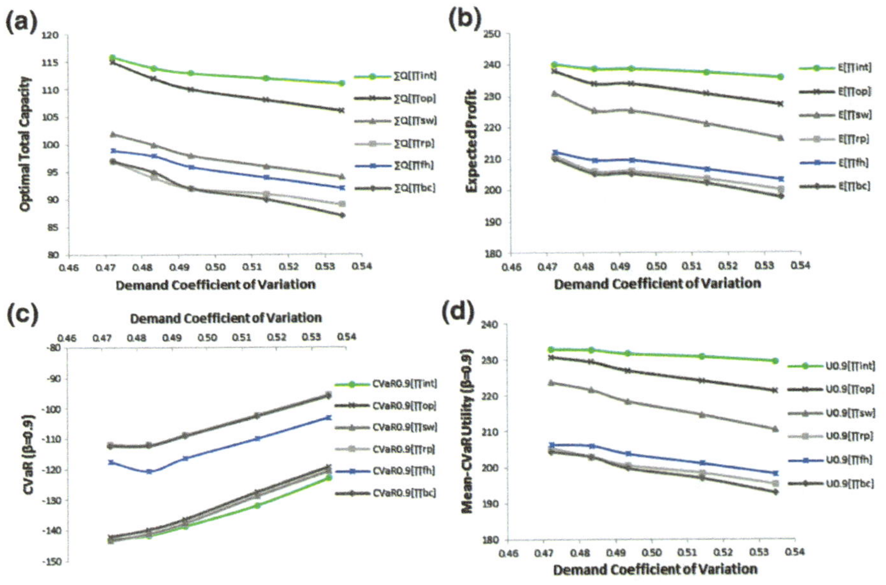

Fig. 5.6 Effect of demand volatility on **a** optimal capacity; **b** expected profit; **c** CVaR; and **d** mean-CVaR

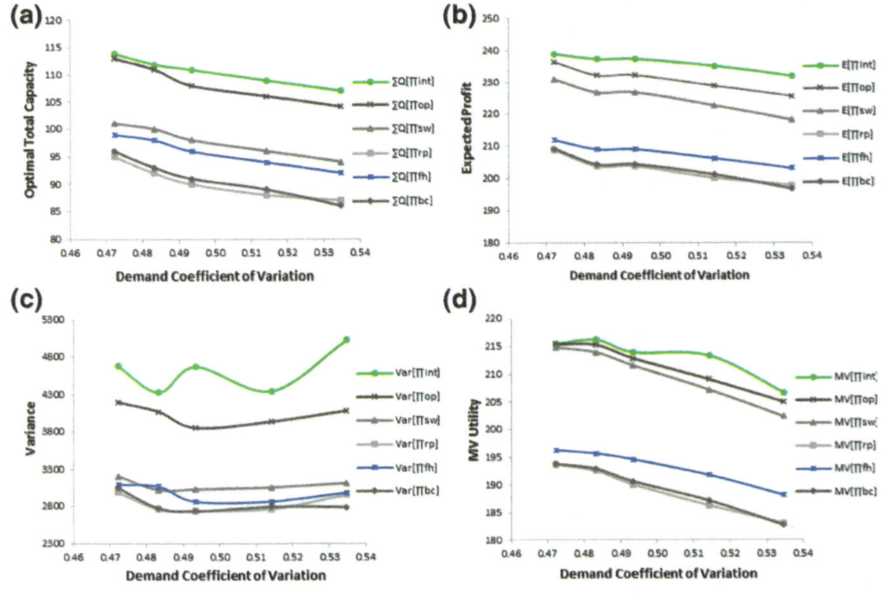

Fig. 5.7 Effect of demand volatility on **a** optimal capacity; **b** expected profit; **c** variance; and **d** MV utility

demonstrates that, with either financial hedging only or integrated risk management, variance is *nonmonotonic* in demand volatility. This observation differs from that in Fig. 5.6c, where CVaR is *increasing* in both financial hedging only and integrated risk management; the difference again stems mainly from our use of CVaR to measure downside risk. Furthermore, in the MV model with two capacities (Ding et al. 2007), profit variance is *nearly constant* (w.r.t. demand volatility) under financial hedging only or integrated risk management. That difference reflects our addition of reliable domestic capacity and its interaction with other hedge options.

5.5 Integrated Hedging with Capacity Reshoring

This chapter explores how a global firm employs capacity reshoring, production switching, and financial hedging to mitigate supply uncertainty, demand risk, and exchange rate fluctuations. We adopt mean-CVaR optimization to decompose operations and finance: operational flexibility maximizes expected profit subject to a CVaR constraint, while financial hedging minimizes CVaR subject to a minimum expected profit. Our main findings are as follows. First, operational flexibility and financial hedging can be complements. Operational flexibility enhances expected profit and reduces downside risk, whereas financial hedging minimizes downside risk and can alter feasibility of capacity portfolios (only by relaxing the CVaR constraint). Second, operational flexibility and financial hedging are substitutes in risk reduction. Financial hedging has greater risk reduction effects in CVaR when used alone. Third, collaboration between operations and finance is indispensable for reducing substitution effects in mean-CVaR utility. Efficient financial hedging depends on the precise estimation of cash flow distribution as determined by operational flexibility, and the CVaR constraint renders the feasibility of capacity choices dependent on financial hedging.

The examination of integrated risk management suggests several future research directions. First, our one-period, two-stage program could be extended to accommodate multiperiod settings in which dynamic financial hedging is adopted concurrently with long-term capacity investments. Second, since our model addresses the case of two currency zones, it might be extended to multiple currencies. Third, we have focused on a single firm's production network; hence a decentralized supply chain could be examined. Fourth, the global firm's pricing decisions (e.g., an exchange rate pass-through) could be incorporated under various competitive settings. Finally, research can be conducted to test empirically the relative effectiveness of operational flexibility and financial hedging; for example, Kim et al. (2006) investigate the relationship of these strategies in mitigating currency risk.

References

Aabo, T., & Simkins, B. J. (2005). Interaction between real options and financial hedging: Fact or fiction in managerial decision-making. *Review of Financial Economics, 14,* 353–369.

Acerbi, C., & Tasche, D. (2002). On the coherence of expected shortfall. *Journal of Banking and Finance, 26,* 1487–1503.

Artzner, P., Delbaen, F., Eber, J., & Heath, D. (1999). Coherent measures of risk. *Mathematical Finance, 9,* 203–228.

Chen, X., Sim, M., Simchi-Levi, D., & Sun, P. (2007). Risk aversion in inventory management. *Operations Research, 55,* 828–842.

Chen, A., Fabozzi, F., & Huang, D. (2012). Portfolio revision under mean-variance and mean-CVaR with transaction costs. *Review of Quantitative Finance and Accounting, 39,* 509–526.

Chen, L., Li, S., & Wang, L. (2014). Capacity planning with financial and operational hedging in low-cost countries. *Production and Operations Management, 23,* 1495–1510.

Choi, S., Ruszczyski, A., & Zhao, Y. (2011). A multiproduct risk-averse newsvendor with law-invariant coherent measures of risk. *Operations Research, 59,* 346–364.

Chowdhry, B., & Howe, J. T. B. (1999). Corporate risk management for multinational corporations: Financial and operational hedging policies. *European Finance Review, 2,* 229–246.

Cohen, M., & Lee, H. (2015). Global supply chain benchmark study: An analysis of sourcing and re-structuring decisions. *Supply Chain Navigator.*

Daimler. (2014). *Mercedes-benz cars at a glance.* Edition 2014.

Daimler. (2016). *Daimler annual report 2015.*

de Grauwe, P. (1988). Exchange rate variability and the slowdown in the growth of international trade. *IMF Staff Papers, 35,* 63–84.

de Treville, S., & Trigeorgis, L. (2010). It may be cheaper to manufacture at home. *Harvard Business Review.* 84–87.

Ding, Q., Dong, L., & Kouvelis, P. (2007). On the integration of production and financial hedging decisions in global markets. *Operations Research, 55,* 470–489.

European Commission. (2009). *The economics of climate change adaptation in EU coastal areas. Country report Germany.*

Financial Times. (2016). *Robot advances help Adidas retrace footsteps.* June 9, 2016.

Ford. (2016). *Delivering profitable growth for all. Ford motor company 2015 annual report.*

Froot, K. A., Scharfstein, D. S., & Stein, J. C. (1993). Risk management: Coordinating corporate investment and financing policies. *Journal of Finance, 48,* 1629–1658.

Gamba, A., & Triantis, A. J. (2014). Corporate risk management: Integrating liquidity, hedging, and operating policies. *Management Science, 60,* 246–264.

GM. (2016). *General motors company 2015 annual report.*

Gotoh, J., & Takano, Y. (2007). Newsvendor solutions via conditional-value-at-risk minimization. *European Journal of Operational Research, 179,* 80–96.

Hommel, U. (2003). Financial versus operative hedging of currency risk. *Global Finance Journal, 14,* 1–18.

Huchzermeier, A., & Cohen, M. A. (1996). Valuing operational flexibility under exchange rate risk. *Operations Research, 44,* 100–113.

Iyengar, G., & Ma, A. K. C. (2013). Fast gradient descent method for mean-CVaR optimization. *Annals of Operations Research, 205,* 203–212.

Kazaz, B., Dada, M., & Moskowitz, H. (2005). Global production planning under exchange-rate uncertainty. *Management Science, 51,* 1101–1119.

Kim, Y., Mathur, I., & Nam, J. (2006). Is operational hedging a substitute for or a complement to financial hedging? *Journal of Corporate Finance, 12,* 834–853.

Kogut, B., & Kulatilaka, N. (1994). Operating flexibility, global manufacturing, and the option value of a multinational network. *Management Science, 40,* 123–139.

Krokhmal, P., Palmquist, J., & Uryasev, S. (2002). Portfolio optimization with conditional value-at-risk objective and constraints. *Journal of Risk, 4,* 43–68.

Mello, A. S., Parsons, J. E., & Triantis, A. J. (1995). An integrated model of multinational flexibility and financial hedging. *Journal of International Economics, 39,* 27–52.

Mello, A. S., Parsons, J. E., & Triantis, A. J. (1996). Flexibility or hedging? *Risk, 9,* 18–19.

Rockafellar, R. T., & Uryasev, S. (2000). Optimization of conditional value.at-risk. *Journal of Risk, 2,* 21–41.

Rockafellar, R. T., & Uryasev, S. (2002). Conditional value-at-risk for general loss distributions. *Journal of Banking and Finance, 26,* 1443–1471.

Sarykalin, S., Serraino, G., & Uryasev, S. (2008). VaR vs. CVaR in risk management and optimization. *Tutorials in Operations Research INFORMS, 2008,* 270–294.

Sauter, M. B., & Stebbins, S. (2016). Manufacturers bringing the most jobs back to America. 24/7 Wall St. April 19, 2016.

Selko, A. (2016). Companies bringing manufacturing jobs back to US. *Industry Week*. August 25, 2016.

Smith, C. W., & Stulz, R. M. (1985). The determinants of firms' hedging policies. *Journal of Financial and Quantitative Analysis, 20,* 391–405.

The Economist. (2009). Managing supply-chain risk for reward. *The Economist Intelligence Unit.*

The Economist. (2012). Floods in Beijing: Under water and under fire. July 24, 2012.

Dada, M., Petruzzi, N. C., & Schwarz, L. B. (2007). A newsvendor's procurement problem when suppliers are unreliable. *Manufacturing and Service Operations Management, 9,* 9–32.

Tomlin, B. (2006). On the value of mitigation and contingency strategies for managing supply chain disruption risks. *Management Science, 52,* 639–657.

Tomlin, B., & Wang, Y. (2005). On the value of mix flexibility and dual sourcing in unreliable newsvendor networks. *Manufacturing and Service Operations Management, 7,* 37–57.

van Mieghem, J. A. (2003). Capacity management, investment and hedging: Review and recent developments. *Manufacturing and Service Operations Management, 5,* 269–302.

van Mieghem, J. A., & Dada, M. (1999). Price versus production postponement: Capacity and competition. *Management Science, 45,* 1631–1649.

Chapter 6
Supply Chain Finance

This chapter offers a theoretical framework for SCF (supply chain finance) and summarizes its applications. We start by proposing a framework for categorizing supply chain finance instruments and then discuss their underlying mechanisms. Next we explore the concept and empirical investigation of working capital management. Finally, we show how emerging conceptual, analytical, and empirical research has established the value—and a foundation for development—of supply chain finance.

6.1 Conceptual Framework of Supply Chain Finance

Although the concept of supply chain finance has been defined in various ways by researchers and practitioners, there are typically three scopes of SCF (Steeman 2014; Liebl et al. 2016; Templar et al. 2016). These are illustrated in Fig. 6.1.

First, "supply chain finance" has been used in reference to the management of monetary flows and financial processes in supply chain. Thus *financial supply chain management* is defined as the "optimized planning, managing, and controlling of supply chain cash flows to facilitate efficient supply chain material flows" (Wuttke et al. 2013a). A closely related definition is that of the *financial supply chain*, or the "network of organizations and banks that coordinate the flow of money and financial transactions via financial processes and shared information systems in order to support and enable the flow of goods and services between trading partners in a product supply chain" (Blackman et al. 2013). Second, SCF can also be viewed as incorporating the set of financial instruments that enhance the efficiency of monetary flows in supply chain: "the use of financial instruments, practices, and technologies to optimize the management of working capital, liquidity, and risk tied up in supply chain processes for collaborating business partners" (Euro Banking Association 2014); or "the inter-company optimization of financing as well as the integration of financing processes with customers, suppliers, and service providers

L. Zhao and A. Huchzermeier, *Supply Chain Finance*, EURO Advanced Tutorials on Operational Research, https://doi.org/10.1007/978-3-319-76663-8_6

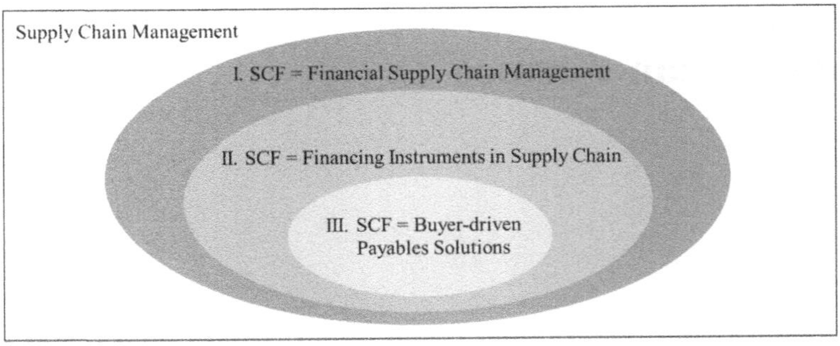

Fig. 6.1 Definitions of supply chain finance

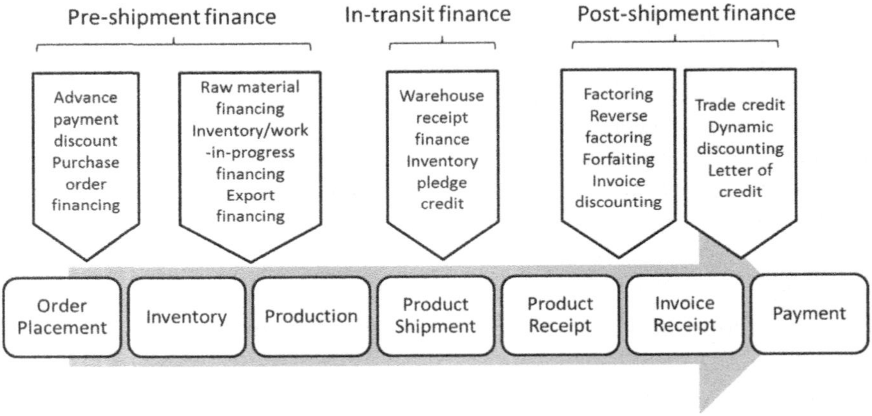

Fig. 6.2 Timing-based classification of supply chain finance

in order to increase the value of all participating companies" (Pfohl and Gomm 2009). Third, SCF may simply denote supplier financing as a buyer-driven payables solution—primarily *reverse factoring*, in which "the lender purchases accounts receivables only from specific informationally transparent, high-quality buyers. The factor only needs to collect credit information and calculate the credit risk for selected buyers, such as large, internationally accredited firms" (Klapper 2006).

The extensive variety of supply chain finance *solutions* can be categorized, from diverse perspectives, in terms of timing of the trigger event, focal point of credit risk, availability of collateral, and financed elements in the balance sheet.

With regard to *trigger event timing*, SCF instruments can be classified into three categories; these are illustrated in Fig. 6.2. (1) Pre-shipment finance enables a supplier to receive funding from a financier—based on a buyer's purchase order—for working capital needs (e.g., the purchase of raw materials, inventory processing, personnel and management costs) before product delivery. Because the collateral for pre-shipment finace is a purchase order instead of an invoice, the credit risk is relatively high; hence the interest rate for advancing liquidity to the supplier is usually high, though it could be reduced in light of a well-established buyer's creditworthiness. An example application of this type of SCF is the launch of a new product; here the supplier needs capital for capacity investment in new production facilities requested by a reputable buyer, which (together with the bank) then initiates financing for the supplier. (2) In-transit finance provides the borrower with a loan from a financial institution, where the loan is based on product or inventory (of a certain quantity and quality) that is currently being transported or enmeshed in other logistics processes. The portable collateral of in-transit finance is the product deposit in shipment, so the associated credit risk is less than in the pre-shipment finance case; hence the loan's interest rate is accordingly somewhat lower. (3) Post-shipment finance establishes a line of credit from a financier for a borrower based on (usually, discounted) accounts receivable. The collateral in this case is the invoice, shipping document, or bill drawn on the buyer. As a consequence, the credit risk is relatively low and the financing rate is favorable.

In terms of the *focal point of credit risk*, Fig. 6.3 shows the three types of SCF instruments (cf. EBA 2014). (1) With accounts receivable finance for supplier, liquidity solutions accelerate the transfer of accounts receivable into cash payments in favor of the supplier by accessing capital from either a financial institution or a buyer. Although this type of SCF instrument finances the supplier's working capital, the interest rate is based on the credit rating of whichever party (supplier or buyer) guarantees the loan. (2) Inventory-related finance enables the lending of capital to either a supplier or a buyer; because the collateral is inventory or a purchase order (i.e., prior to issuance of an invoice), the interest rate is typically higher than on loans collateralized by confirmed invoices. (3) Finally, accounts payable finance for buyer offers an early payment discount and an extended payment period from the supplier—or a guarantee from a financial institution—to ensure payment by the buyer. The credit risk and hence the interest rate depend on the buyer's credit rating.

From the perspective of *availability of collateral*, we identify two categories of supply chain finance (cf. Navas-Alemán et al. 2012); see Fig. 6.4. First,

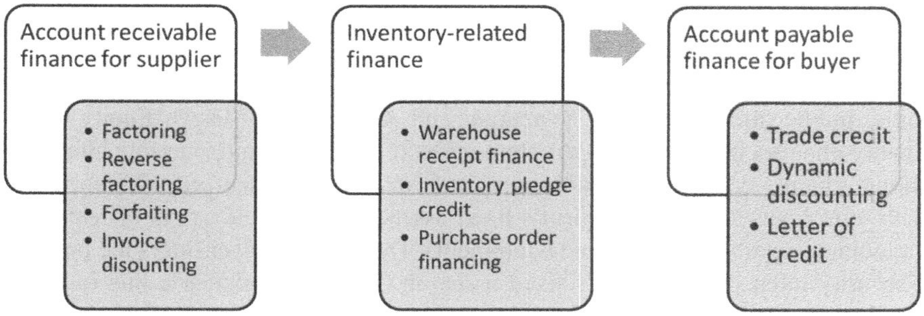

Fig. 6.3 Supply chain finance categorization by focal point of credit risk

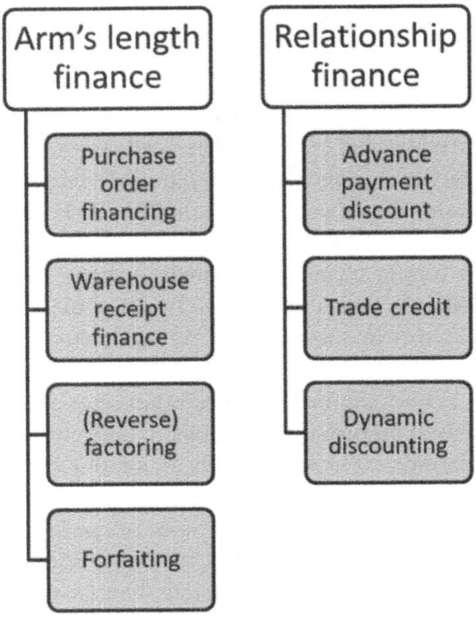

Fig. 6.4 Classification of supply chain finance based on availability of collateral

"arm's-length" finance instruments are established based on verifiable information or tangible collateral (e.g., invoice, bill of exchange, purchase order) and so are associated with enforcement mechanisms in the event of nondelivery or nonpayment. Thus the credit risk of arm's-length finance instruments can be estimated more accurately and controlled by the financial institution based on available information about verifiable collateral. Second "relationship" finance instruments depend on mutual trust or non-binding options in contract/agreement reflecting the relation's transaction history, so there is seldom any tangible collateral. The lender in relationship finance is typically a supply chain member that has information on the borrower's creditworthiness and trade history in a long-term relationship; this background enables the proper evaluation and management of credit risk by the lender.

Finally, as regards a *balance sheet's financed elements*, the three types of SCF instruments are shown in Fig. 6.5. (1) Equity-related finance refers to fund-raising through the sales of shares in—or the ownership transfer of—a firm; it includes a variety of activities that range from raising capital from extant shareholders, to initial public offerings (IPOs), to merger and acquisitions (M&A). Equity-related finance changes the ownership structure of the related firms and therefore affects the supply chain's competition landscape. (2) Fixed-asset finance provides borrowers with term loans secured by a firm's fixed assets (i.e., tangible assets or real estate including property, facilities, and equipment). Thus the borrower offers the financier a security interest in the firm's fixed assets in order to guarantee a line of credit. (3) Working capital finance aims to fund a firm's daily operations (rather than

Fig. 6.5 Types of supply chain finance by financed elements in the balance sheet

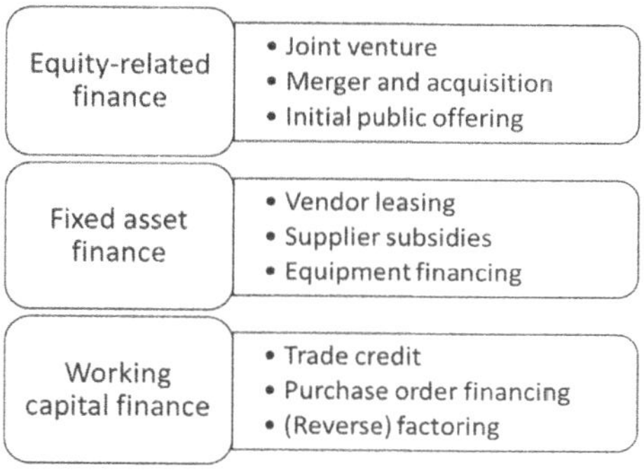

long-term assets or investments), such as accounts payable for raw materials and personnel costs, to ensure sufficient liquidity in the supply chain.

6.2 Supply Chain Finance Instruments

In this section we review the supply chain finance instruments frequently employed in practice. These instruments can be distinguished by the timing of funding, collateral, beneficiary, and credit guarantee provider; see Table 6.1.

Under *advance payment discount*, a buyer finances the supplier by advancing payment at a discount prior to product shipment; this is also known as "cash in advance". The unit discount offered by the supplier incentivizes the buyer to make an advance payment, which can be used to fulfill the supplier's working capital needs. The advance payment can also alleviate the supplier's budget constraint and thereby mitigate the buyer's risk of procurement shortage.

Purchase order financing is a scheme whereby small and medium-sized enterprise (SME) suppliers receive funding, prior to product delivery, from a financial institution based on a reputable buyer's (discounted) purchase order. In the absence of a guarantee from the buyer, interest rates for this type of financing depend on the supplier's credit rating. When the purchase order financing loan is secured by a reputable buyer, the supplier's financing rate depends on the buyer's creditworthiness; this setup is called *buyer-backed* purchase order financing. This variant allows an SME supplier to contract for a larger order quantity given the external funding provided—thanks to the buyer's credit rating—by a bank.

Under *warehouse receipt finance*, a financial institution loans funds to a supplier based on a warehouse receipt that certifies—as portable collateral—the secured storage of product in a specified quantity and quality. Figure 6.6 depicts the sequence of events in a warehouse receipt finance scheme; here the transfer of a warehouse receipt from supplier to financier conveys the right to withdraw a certain

Table 6.1 Overview of supply chain finance instruments

SCF instrument	Timing of funding	Collateral	Credit guarantee provider	Financier	Beneficiary	Description
Advance payment discount	Pre-shipment	Purchase order	Buyer	Buyer	Supplier and buyer	Buyer pays supplier at unit discount before shipment
Purchase order financing	Pre-shipment	Purchase order	Supplier	Bank	Supplier	Bank buys supplier's receivables based on purchase order before shipment
Buyer-backed purchase order financing	Pre-shipment	Validated purchase order	Buyer	Bank	Supplier	Bank buys supplier's receivables guaranteed by buyer based on purchase order before shipment
Warehouse receipt finance	In-transit	Warehouse receipt	Warehouse	Bank	Supplier	Bank offers funding to supplier based on warehouse receipt
Inventory pledge finance	In-transit	Pledged inventory	Borrower	Bank	Borrower	Bank offers funding to borrower secured by pledged inventory
Trade credit	Post-shipment	Invoice	Supplier	Supplier	Buyer	Supplier offers buyer early payment discount; charges interest on extended payment
Dynamic discounting	Post-shipment	Invoice	Supplier	Supplier	Buyer	Supplier offers buyer discount based on length of time until payment is received
Recourse factoring	Post-shipment	Invoice	Supplier	Bank	Supplier	Supplier sells receivables to bank with recourse
Non-recourse factoring	Post-shipment	Invoice	Supplier and buyer	Bank	Supplier	Supplier sells receivables to bank without recourse
Reverse factoring	Post-shipment	Validated invoice	Buyer	Bank	Supplier	Supplier sells receivables to bank guaranteed by buyer
Forfaiting	Post-shipment	Invoice	Exporter	Bank	Importer	Exporter sells receivables to bank
Letter of credit	Post-shipment	Bill of lading	Buyer	Bank	Supplier	Bank guarantee for buyer's payment to supplier

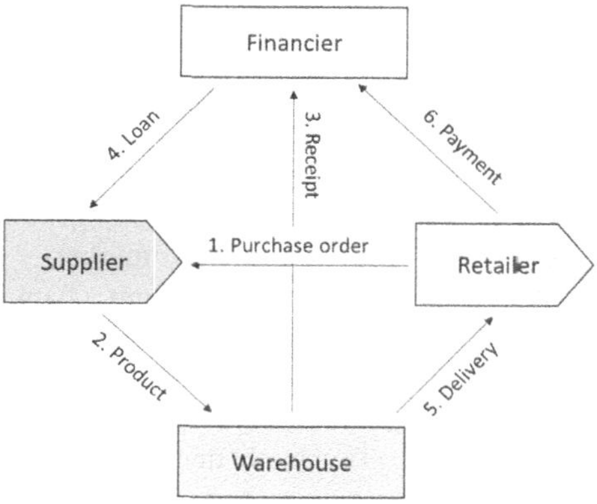

Fig. 6.6 Warehouse receipt finance scheme

amount of the commodity, at any time, from the secured warehouse. The financier provides a loan up to an agreed percentage (the discounted value) of the stored product.

Inventory pledge finance is a closely related form of funding that is provided by a financial institution to a borrower while using secured inventory as collateral. This form of finance can be used to fulfill working capital needs for capacity expansion, equipment renewal, or material supply. When other types of firm assets are already leveraged, pledged inventory can serve as collateral to secure a loan.

Trade credit refers to a contract term whereby a buyer receives a discount on the wholesale price when payment is remitted within a specified time; a buyer that remits later must pay, in addition to the (discounted) wholesale price, a pre-specified interest payment. Trade credit is one of the most often used short-term financing instruments in global trade (Rajan and Zingales 1998; Giannetti et al. 2011).

Dynamic discounting, which is based on trade credit, amounts to a discount on the wholesale price that decreases gradually over time—unlike a fixed discount rate for a certain number of days and then no discount afterwards. This SCF instrument enables the buyer to receive a slightly lower discount rate after the early payment period specified in a trade credit contract.

Factoring refers to the practice whereby a supplier receives a line of credit from a financial institution by selling accounts receivable from a buyer at a discount for immediate payment. In *recourse* factoring, the financier (factor) has the right to require payment from the supplier for any unpaid invoice amounts; in this case, the interest rate depends entirely on the supplier's creditworthiness. In *non-recourse* factoring, the factor assumes the buyer's risk of nonpayment; hence the interest rate depends on both the supplier's and the buyer's credit ratings.

Reverse factoring is a funding scheme initiated by a reputable buyer to provide a guarantee for the transfer of a supplier's accounts receivable to a financial institution. Factoring is more prevalent in developed economies, where suppliers typically have higher credit ratings; reverse factoring enables SME suppliers to obtain financing at a more favorable interest rate (than would otherwise be possible) because they are backed by a reputable buyer's consolidated invoice.

Forfaiting enables an exporter to sell accounts receivable from an importer at a discount to a financial institution (the forfaiter). In forfaiting "without recourse", the financial institution bears the default risk of the importer's payment. This SCF instrument can transfer an exporting firm's accounts receivable as a debt instrument that is tradable on a secondary market.

A *letter of credit* is a letter from a bank to a supplier guaranteeing that a buyer's payment will follow, in a specified amount and on a certain date, upon the delivery of certain documents. A letter of credit may be transferrable—that is, the beneficiary (supplier) can assign, upon mutual agreement, another firm the right to draw on that credit. The bank that issues a letter of credit typically requires pledged securities or documents (e.g., a bill of lading) for collateral.

6.3 Working Capital Management in Supply Chains

Working capital management can be studied from both the finance and the supply chain perspective. From a financial viewpoint, (net) working capital is defined as current assets *minus* current liabilities. From a supply chain viewpoint, working capital amounts to inventories, *plus* accounts receivable, *minus* accounts payable. The performance of working capital is measured by the cash conversion cycle (CCC)—otherwise known as the cash-to-cash (C2C) cycle—which is defined as "the interval between the time cash expenditures are made to purchase inventory for use in the production process and the time that funds are received from the sale of the finished product. This time interval is measured in days and is equal to the net of the average age of the inventory plus the average collection period minus the average age of accounts payable" (Schilling 1996).

Thus the cash conversion cycle can be calculated as follows:

$$\text{CCC} = \text{Days inventory outstanding (DIO)} + \text{Days accounts receivable outstanding (DSO)} \\ - \text{Days accounts payable outstanding (DPO)}.$$

The three components of CCC are defined as follows:

$$\text{DIO} = (\text{Inventory} \times 365)/\text{Cost of sales},$$
$$\text{DSO} = (\text{Accounts receivable} \times 365)/\text{Revenue},$$
$$\text{DPO} = (\text{Accounts payable} \times 365)/\text{Total cash operating expenses}.$$

With respect to the timeline of events, the relationship among CCC components is illustrated in Fig. 6.7.

In practice, a *positive* CCC value indicates the number of days between a firm's billing accounts payable and collecting accounts receivable whereas a *negative*

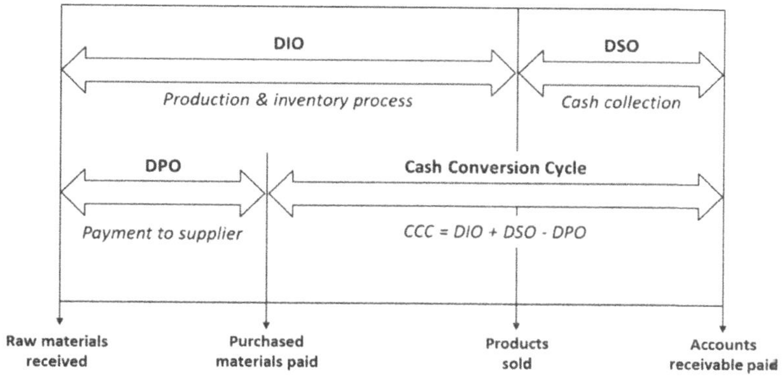

Fig. 6.7 Cash conversion cycle. Adapted from (Dahiya 2012)

CCC value indicates the number of days between a firm receiving payment from customers and remitting to suppliers. A firm's working capital performance, as measured by CCC, is often one of its key performance indicators.

There is usually a negative correlation between the firm's CCC value and its profitability; the latter can be measured by several related KPIs, which include economic value added (EVA®)[1] and return on capital employed (ROCE). The relationship between EVA and working capital management is shown in Fig. 6.8, where NOA denotes Net Operating Assets and where A/R and A/P abbreviate (respectively) accounts receivable and accounts payable. Economic value added is defined as the net operating profit after taxes (NOPAT) *less* the cost of capital of both equity and debt. A firm's efforts to improve working capital management—which is measured by CCC—may include reducing inventories and accounts receivable as well as increasing accounts payable. That strategy leads to a reduction in both current and fixed assets and to a higher credit rating; as a result, the firm's weighted average costs of capital (WACC) are lower. One benefit of the consequent better working capital position is that a firm can release cash flow originally tied up in daily operations, thus inducing (indirectly) a fall in production costs and thereby a rise in operating income (cf. Hofmann and Belin 2011).

6.4 Emerging Research on Supply Chain Finance

The study of supply chain finance began with conceptual research and frameworks. Figure 6.9 presents such a framework, one that incorporates the involved partners, the relevant components, and the interconnections among material, monetary, and information flows. The involved partners are suppliers who sell products to buyers,

[1]Economic value added (EVA®) is a registered trademark of Stern Stewart Management.

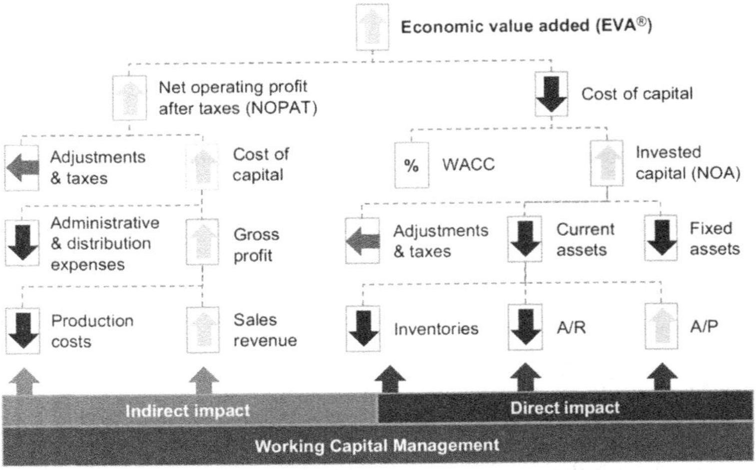

Fig. 6.8 Relationship between economic value added and working capital management. *Source* Hofmann and Belin (2011)

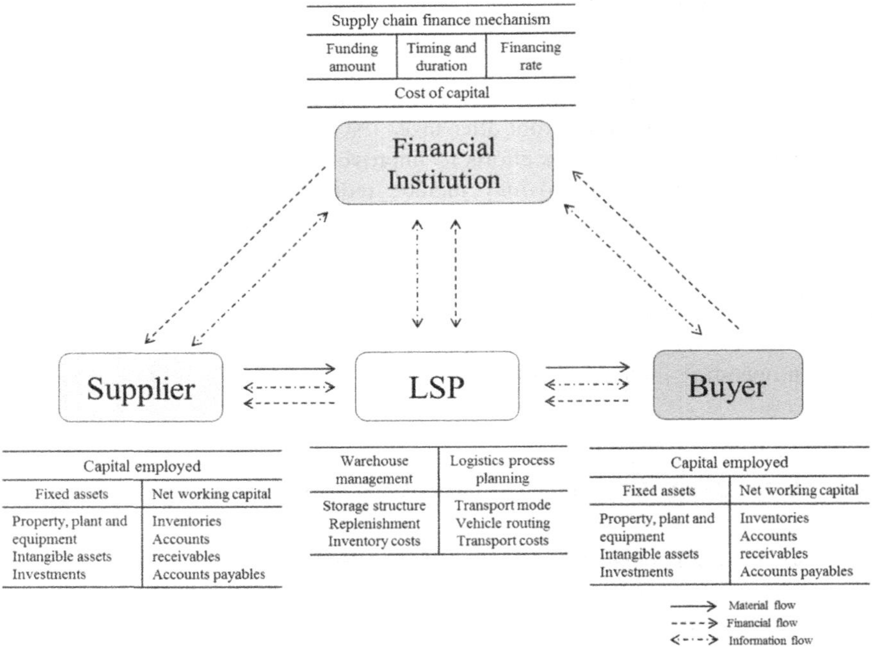

Fig. 6.9 A framework for supply chain finance

where transactions involving monetary flows are intermediated by a financial institution while those involving material flows are intermediated by a logistics service provider (LSP). The adoption of supply chain finance instruments affects the capital that is employed—in other words, the fixed assets and net working capital of both supplier and buyer.

The mechanism of a chosen SCF instrument determines timing of the trigger event, the duration and amount of funding, and the financing rate based on the focal point of credit risk, which combine to establish the cost of capital for the financed party. An LSP's traditional roles are to manage warehouse inventory and replenishment and to optimize the logistics process by selecting the mode of transportation and vehicle routing. Yet an LSP could also be a party to supply chain finance by providing, as collateral for a loan, product inventory based on a warehouse receipt (see Fig. 6.6) or by providing financial services in lieu of a financial institution (e.g., PCH International serves as an intermediary for financing the Chinese suppliers of Western buyers).

As various SCF instruments are gradually adopted in practice, the investigation of SCF evidence has proceeded to explore its drivers, major benefits, potential resistance, adoption process, and overall effects on supplier–buyer relationships. Table 6.2 provides a summary of this research. Supply chain finance programs are typically initiated by established buyers (or suppliers), financial institutions (banks), or specialized service providers (such as LSPs) to provide financial assistance for SME suppliers (or buyers) in need of working capital. Thus working capital positions in the supply chain are the primary antecedents of SCF adoption and also determine the types of SCF instruments used.

The major benefits of SCF programs rely on the reduction of financing costs for suppliers (due to the interest spread between SMEs and established firms) or for buyers (because of extended payment terms). These programs have the further advantage of strengthening supply chain relationships, increasing its members' negotiating power, and improving service (cf. Lekkakos and Serrano 2016). The dependence between supplier and buyer—and their respective bargaining power—are determined by the buyer's order quantity, the ordered product's strategic value, and the intensity of market competition. All of these factors affect, in turn, the product's purchase price (Liebl et al. 2016). The SCF solutions driven by incentives to improve the adopter's own financial performance are typically implemented based on bargaining power, and SCF practices driven by incentives to secure the entire supply chain through risk mitigation efforts depend on a high level of trade process digitalization (Caniato et al. 2016).

It is worth noting that the transmission of data in SCF transactions must comply with the involved countries' applicable electronic security laws. For example, the Nacional Financiera (Nafin) development bank in Mexico facilitates SCF services to SME suppliers supported by Internet-based management information systems (MIS), which are implemented within the framework of the Law of Conservation of Electronic Documents, the Electronic Signature Law, and the Federation fiscal code (which includes digital certification standards). The Nafin program provides several SCF products—which include factoring, reverse factoring, and purchase order

Table 6.2 Summary of supply chain finance research

Research method	Sample size/scope of model	SCF instruments	Examined factors	References
Conceptual research	Characterization and elements of supply chain finance	Supply chain finance framework	Institutional actors, collaboration characteristics, functional perspective of SCF	Hofmann (2005)
Model, conceptual research	A borrower and an investor in a supply chain and an investor in a financial market	Internal and external financing in supply chain	Objects, actors, and levers of supply chain finance, cash conversion cycle, costs of capital	Pfohl and Gomm (2009)
Conceptual research, case study, numerical example	Supplier, receiver, logistics service provider, financial service provider	Inventory financing	Conflict of goals among various parties, additional profit from financing activities, value of financed goods	Hofmann (2009)
Conceptual research	Automotive supply chain	Natural hedging, financial hedging, supplier financing	Currency and commodity price risks, centralization of commodity purchasing, transportation costs	Hofmann (2011)
Case study	6 cases of European firms	Adoption of supply chain finance	Redefining, restructuring, supplier involvement, dissemination, relationship strength, SCF leverage	Wuttke et al. (2013a)
Case study	8 cases based on 40 interviews with European firms	Buyer credit, inventory/ work-in-progress financing, reverse factoring, letters of credit, open account credit, bank loan	Financial supply chain management practice and performance impact, interdepartmental interfaces, supplier relationships	Wuttke et al. (2013b)
Case study	Motorola	Global financial supply chain strategy	Quality measure of financial processes, financial and banking information systems, cash management strategy	Blackman et al. (2013)
Case study	28 interviews in 11 case studies on firms from Europe, the US, and China	Reverse factoring	Objectives and antecedents of implementation, barriers	Liebl et al. (2016)

(continued)

Table 6.2 (continued)

Research method	Sample size/scope of model	SCF instruments	Examined factors	References
Case study	14 firms in Italy	(Reverse) factoring, inventory financing, dynamic discounting, invoice auction, VMI, consignment stock	Collaboration, bargaining power, level of digitalization, financial attractiveness	Caniato et al. (2016)
Survey	145 responses from Swiss companies	Financing of buyer–supplier dyads	Buyer–supplier information sharing, financing alignment and performance, corporate financing strategy alignment	Wandfluh et al. (2015)
Model, simulation	A set of buyer-centric supply chains and a bank	Adoption of reverse factoring	Market competition, supply chain receivables, interest rates, supplier's working capital goals	Iacono et al. (2014)
Model, simulation	A SME supplier sells to a buyer	Reverse factoring	Demand variability, profit margin, access to external financing	Lekkakos and Serrano (2016)
Model, simulation	Supplier and buyer	Adoption of supply chain finance	Global and local exposure, extension of payment terms, buyer's introduction timing	Wuttke et al. (2016)
Model, simulation	A buyer and multiple suppliers	Dynamic discounting	Daily discount, value of liquidity, net operating working capital	Gelsomino et al. (2016)

financing—to Mexican SME suppliers. More than 98% of these products operate electronically to reduce management costs and enhance accessibility (Klapper 2006).

The main costs of establishing SCF programs are (i) management costs of inter-organizational supply chain collaboration and of intra-firm cross-functional coordination and (ii) investments in digital platforms for the trade process (cf. Wandfluh et al. 2015). For instance, the Swiss Post Group offers combined logistics and financial services in a pilot project with Procter & Gamble (P&G). Since the retailers of P&G products in Switzerland vary considerably in size, lower credit ratings of SME retailers could lead to higher capital costs. In this case, the order management is bundled in a centralized logistics platform from which Swiss Post Logistics (the LSP) provides joint logistics and financing services as a supply chain intermediary. Swiss Post Logistics relies on information systems to serve as the bulk (wholesale) buyer of P&G products and as the provider of financial services (e.g., invoicing, creditor management, debt collection) for retailers while incurring lower capital costs than they would individually. Under this arrangement, then, Swiss Post Logistics not only takes ownership of goods as an LSP but also reduces capital costs in the supply chain by receiving quantity discounts as a financial service provider. Thus Swiss Post Logistics can sell goods to retailers at standard prices determined by P&G's policies *plus* a logistics and finance fee (Hofmann 2009). In recent years, innovations in financial technology (FinTech) have emerged rapidly and are expected to gain momentum continuously (EY 2017). Disruptive innovations such as blockchain technology (which features peer-to-peer network, cryptography and digital signature, immutability of data, and decentralized consensus, see Antonopoulus 2015) can enable and enhance SCF adoption in financial services.

The success of supply chain finance instruments depends crucially on the resulting allocation of benefits among supply chain parties. It is recommended that buyers with relatively higher credit ratings, ordering higher procurement quantities, and offering longer payment terms adopt SCF at an early stage because they have the most to gain from modern SCF instruments. In addition, working capital goals should be specified to align incentives in supply chain collaboration and cross-functional coordination. The SCF team needs to work closely with managers from the procurement, operations, IT, legal, treasury, and finance departments. Moreover, the extent of digitalization plays a key role in providing the real-time transparency of supplier invoice processing and other functions. Ensuring that a corporation's enterprise resource planning (ERP) system is compatible with the SCF platform typically requires both managerial effort and technical modifications. Buyers can categorize suppliers based on their strategic importance and creditworthiness, bringing the most crucial ones onboard first and then gradually incorporating more suppliers into the system. Even though an SCF program benefits SME suppliers the most, in practice it is larger suppliers that first join the pilot programs (in anticipation of those programs' off-balance sheet effects). The SME suppliers join later—following the initial success of SCF adoption (Wuttke et al. 2013a, 2016).

References

Antonopoulus, A. M. (2015). *Mastering Bitcoin: Unlocking digital cryptocurrencies* (1st ed.). Sebastopol: O'Reilly Media.

Blackman, I. D., Holland, C. P., & Westcott, T. (2013). Motorola's global financial supply chain strategy. *Supply Chain Management: An International Journal, 18,* 132–147.

Caniato, F., Gelsomino, L. M., Perego, A., & Ronchi, S. (2016). Does finance solve the supply chain financing problem? *Supply Chain Management: An International Journal, 21,* 534–549.

Dahiya, S. (2012). *Working capital simulation: Managing growth.* Harvard Business School Case.

Euro Banking Association [EBA]. (2014). *Supply chain finance—EBA European market guide.*

Ernst & Young [EY]. (2017). *EY FinTech adoption index 2017: The rapid emergence of FinTech.*

Gelsomino, L. M., Mangiaracina, R., Perego, A., & Tumino, A. (2016). Supply chain finance: A literature review. *International Journal of Physical Distribution & Logistics Management, 46,* 348–366.

Giannetti, M., Burkart, M., & Ellingsen, T. (2011). What you sell is what you lend? Explaining trade credit contracts. *The Review of Financial Studies, 24,* 1261–1298.

Hofmann, E. (2005). Supply Chain Finance: Some conceptual insights. In Rainer Lasch & Christian G. Janker (Eds.), *Logistik management—Innovative logistikkonzepte* (pp. 203–214). Wiesbaden: Deutscher Universitätsverlag.

Hofmann, E. (2009). Inventory financing in supply chains: A logistics service provider-approach. *International Journal of Physical Distribution & Logistics Management, 39,* 716–740

Hofmann, E. (2011). Natural hedging as a risk prophylaxis and supplier financing instrument in automotive supply chains. *Supply Chain Management: An International Journal, 16,* 128–141.

Hofmann, E., & Belin, O. (2011). *Supply chain finance solutions.* Springer briefs in business Berlin: Springer.

Iacono U.D., Reindorp M., & Dellaert N. (2014). Market adoption of reverse factoring. *International Journal of Physical Distribution & Logistics Management, 45,* 286–308

Klapper, L. (2006). The role of factoring for financing small and medium enterprises. *Journal of Banking Finance, 30,* 3111–3130.

Lekkakos, S. D., & Serrano, A. (2016). Supply chain finance for small and medium sized enterprises: The case of reverse factoring. *International Journal of Physical Distribution & Logistics Management, 46,* 367–392.

Liebl, J., Hartmann, E., & Feisel, E. (2016). Reverse factoring in the supply chain: Objectives, antecedents and implementation barriers. *International Journal of Physical Distribution & Logistics Management, 46,* 393–413.

Navas-Alemán, L., Pietrobelli, C., & Kamiya, M. (2012). *Inter-firm linkages and finance in value chains.* Working paper, Inter-American Development Bank.

Pfohl, H.-C., & Gomm, M. (2009). Supply chain finance—Optimizing financial flows in supply chains. *Logistics Research, 1,* 149–161.

Rajan, R. G., & Zingales L. (1998). Financial dependence and growth. *American Economic Review, 88,* 559–586.

Schilling, G. (1996). Working capital's role in maintaining corporate liquidity. *TMA Journal, 16,* 4–7.

Steeman, M. (2014). The power of supply chain finance: How companies can apply collaborative finance models in their supply chain to mitigate risks and reduce costs. Windesheim University of Applied Sciences.

Templar, S., Findlay, C., Hofmann, E. (2016). *Financing the end-to-end supply chain: A reference guide to supply chain finance.* KoganPage.

Wandfluh, M., Hofmann, E., & Schoensleben, P. (2015). Financing buyer–supplier dyads: an empirical analysis on financial collaboration in the supply chain. *International Journal of Logistics Research and Applications, 19,* 1–16.

Wuttke, D. A., Blome, C., Foerstl, K., & Henke, M. (2013a). Managing the innovation adoption of supply chain finance—Empirical evidence from six European case studies. *Journal of Business Logistics, 34,* 148–166.

Wuttke, D. A., Blome, C., Heese, H. S., & Protopappa-Sieke, M. (2016). Supply chain finance: Optimal introduction and adoption decisions. *International Journal of Production Economics, 178,* 72–81.

Wuttke, D. A., Blome, C., & Henke, M. (2013b). Focusing the financial flow of supply chains: An empirical investigation of financial supply chain management. *International Journal of Production Economics, 145,* 773–789.

Chapter 7
Managing Supplier Financial Risk with Pre-shipment Finance Instruments

In this chapter, we present a stylized model for managing supplier financial distress risk by two pre-shipment finance instruments: advance payment discount (APD) and buyer-backed purchase order finance (BPOF). We analyze the mechanisms of APD and BPOF, respectively. Next, we characterize the equilibria between APD and BPOF in both single-financing and dual-financing schemes. Moreover, we show numerically the effects of demand variability and the retailer's internal capital level on the financing equilibrium and supply chain efficiency. We conclude with a brief summary and directions for future research.

7.1 Trade Finance Instruments

Financial distress of suppliers can strongly affect supply chain efficiency. For example, in February 2008, Chrysler temporarily closed four assembly plants and canceled one shift at a fifth plant—citing a parts shortage after its supplier, Plastech, filed for Chap. 11 bankruptcy protection (Nussel and Sherefkin 2008). A buyer can alleviate its supplier's financial distress via trade finance instruments (Chauffour and Malouche 2011). In particular, a buyer can finance a supplier by advancing payment at a discount prior to product shipment; this procedure is referred to as "cash in advance".[1] Figure 7.1 shows a 2008 International Monetary Fund estimate on the market share of trade finance instruments, where advance payment discount (APD) represents 19–22% (i.e., 3–3.5 trillion US dollars) of global trade finance. In the literature on supply chain finance, bank finance and trade credit (i.e , open account) have been frequently addressed (Jing et al. 2012; Kouvelis and Zhao 2012; Cai et al. 2014). In contrast, there is a dearth of research on advance payment

[1]The only (minor) difference between *cash in advance* and *advance payment discount* is whether the supply contract includes a unit discount. Thus APD can be seen as the special case of cash in advance in which the buyer receives such a discount.

L. Zhao and A. Huchzermeier, *Supply Chain Finance*, EURO Advanced Tutorials on Operational Research, https://doi.org/10.1007/978-3-319-76663-8_7

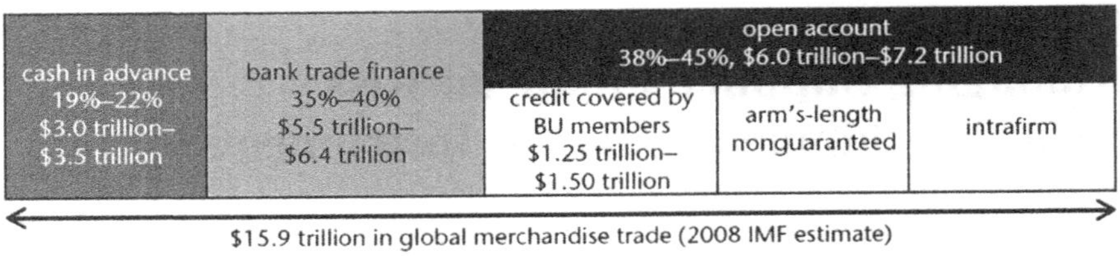

Fig. 7.1 Trade finance arrangements by market share (Chauffour and Malouche 2011; BU Berne Union)

discount despite its prevalence in practice. Hence, this chapter aims to assess the value of APD in managing supplier financial distress.

Another trade finance instrument to mitigate the supplier's capital constraint is buyer-backed purchase order financing (BPOF), whereby a financial institution provides a loan based on a purchase order—guaranteed by the buyer—to fund the supplier before order delivery (Martin 2010; Tice 2010). A program of this type was launched by Nacional Financiera (Nafin), the Mexican state-owned development bank (Klapper 2006). Furthermore, the specialty lender PurchaseOrderFinancing.com has secured more than $750 million since 2002 to boost US, UK, Canadian, and Chinese business growth (PurchaseOrderFinancing.com 2016). However, relatively less attention has been paid to buyer-backed purchase order financing in spite of its emergence as a viable trade finance instrument. Therefore, our research attempts to evaluate the impact of BPOF and its interaction with APD in a budget-constrained supply chain.

In sum, we seek to address the following research questions.

1. Which financing strategy (APD or BPOF) is more efficient and should be chosen by the retailer?
2. What is the financing equilibrium of APD and BPOF when both instruments can be adopted?
3. How do the retailer's optimal sourcing and financing decisions affect supply chain efficiency?

To answer these questions, we first analyze respectively the mechanisms of APD and BPOF in a capital-constrained supply chain of one small- and medium-sized enterprise (SME) supplier and an established retailer. Moreover, we derive the financing equilibrium between APD and BPOF and show that the equilibrium region of APD is increasing in both the retailer's internal capital level and demand variability. Hence, greater demand uncertainty increases the need for risk sharing in the supply chain. When both trade finance instruments are viable, the retailer prefers APD and will initiate BPOF only under certain conditions. That is, the sourcing of capital from the retailer has a higher priority than the borrowing of capital from a financial institution. In addition, we find that there are considerable costs of competition penalty when demand variability and the retailer's internal capital level (respectively) are within certain intervals. Hence, incentive alignment between supply chain partners is crucial for mitigating competition penalty.

7.2 A Supply Chain with APD and BPOF

We consider a supply chain consisting of one SME supplier and one established retailer. All parties are assumed to be risk neutral. The retailer (she) orders q units of a product from the supplier (he), who has capacity K. Both the supplier and retailer operate with limited amounts of internal capital before shipment, thus either may be capital constrained. The retailer and supplier have long-term capital structures financed solely by equity and have short-term debts due before the sales season. Thus both firms face risk of financial distress before shipment, which occurs (or not) depending on whether internal assets are enough to cover loan obligations. The supplier has internal asset A_s and short-term debt L_s. The supplier's asset A_s is assumed to be stochastic and not realized until the sales season (cf. Babich 2010; Hortaçsu et al. 2011; Yang et al. 2015). It has the cumulative distribution function (CDF) $\Phi(A_s)$, probability density function (PDF) $\phi(A_s)$, and $A_s \in \lfloor \underline{A_s}, \overline{A_s} \rfloor$ for $0 \leq \underline{A_s} < \overline{A_s} \leq \infty$. Financial default is associated with a realization of the asset below threshold $\widetilde{A_s}$. The supplier's short-term debt L_s is deterministic and will be due before the sales season.[2] As an established firm, the retailer has a higher credit rating with the financial institution than the SME supplier. Our notation and assumptions are summarized in Table 7.1.

The supplier is capital-constrained, and the retailer could also experience financial distress when she provides financing to the supplier. For instance, both GM and Chrysler have undergone Chap. 11 reorganization in 2009 while offering financial assistance to SME suppliers (Marr 2009). We assume that the retailer's asset and liability are deterministic.[3] The capital market is imperfect. That is, when a firm cannot repay debt obligations, it can either be liquidated or negotiate with creditors while enduring a costly reorganization process. In case of liquidation, the cost of financial default is a proportion $1-\gamma$ $(0<\gamma<1)$ of firm value. In case of reorganization, the cost of financial distress is a proportion $1-\alpha$ $(0<\alpha<1)$ of raised capital (Leland 1994; Gamba and Triantis 2014).

Demand D is stochastic and not realized until the sales season. The demand distribution function $F(D)$ is absolutely continuous with density $f(D)>0$ and support $[a, b)$ for $0 \leq a < b \leq \infty$; it has a finite mean and an inverse $F^{-1}(D)$. The hazard rate $h(D) \triangleq f(D)/\bar{F}(D)$ is increasing in D, where $\bar{F}(D) = 1 - F(D)$. $H(D) \triangleq Dh(D)$ denotes the generalized failure rate; then $H(D)$ is monotonically increasing in D. Suppose that F has a strictly increasing generalized failure rate (IGFR), and $h(D) = Df(D)/(1-F(D))$. Without loss of generality (w.l.o.g.), the

---

[2]In practice, this refers to the case in which the supplier has been informed by debt holders about the amount and due date of liabilities, while the value of internal asset is subject to market variables (e.g., interest rates and commodity prices). Since this leads to the stochastic internal capital level, adding stochasticity to debt will not alter the structure of our main results.

[3]An established firm (large corporation) typically employs financial hedging to mitigate the impact of market dynamics (incl. commodity price, exchange and interest rates). Hence, the value of her asset and liability can be viewed as deterministic in a short term.

Table 7.1 Summary of notation and assumptions

Symbol	Description	Assumptions
π_i	The expected profits of the supplier, retailer, and financial institution	$i = s, r, c^*$
A_i	Asset level of supplier/retailer	Exogenous, $i = s, r$
L_i	Short-term debt of the supplier/retailer due before order delivery	Exogenous, $i = s, r$
D	Product demand	Exogenous, stochastic
$f(\cdot)$	The probability density function (PDF) of demand distribution	Exogenous, $f(D) > 0$
$F(\cdot)$	The cumulative distribution function (CDF) of demand distribution	Exogenous, $F(D)$ has support $[a, b)$ for $0 \leq a < b \leq \infty$
$\bar{F}(\cdot)$	The complementary cumulative distribution function (CCDF) of the demand distribution, $\bar{F}(D) = 1 - F(D)$	Exogenous
$h(\cdot)$	The failure rate of the demand distribution, $h(\cdot) = f(\cdot)/\bar{F}(\cdot)$	Exogenous, increasing in D
$\phi(\cdot)$	The PDF of the supplier's asset distribution	Exogenous, $A_s \in [\underline{A_s}, \overline{A_s}]$ for $0 \leq \underline{A_s} < \overline{A_s} \leq \infty$
$\Phi(\cdot)$	The CDF of the supplier's asset distribution	Exogenous
$\bar{\Phi}(\cdot)$	The CCDF of the supplier's asset distribution, $\bar{\Phi}(A_s) = 1 - \Phi(A_s)$	Exogenous
K	Capacity of the supplier	Decision variable, $K \in \mathbb{R}_+$
q	Order quantity of the retailer to the supplier	Decision variable
w	Unit wholesale price	Decision variable, $p > w > c_p + c_k$
d	Discount rate of the retailer's advance payment to the supplier before order delivery	Decision variable, $d \in [0, \frac{w - c_p - c_k}{w}]$
c_p	Unit production cost of the supplier	Exogenous, deterministic, $w > c_p + c_k$
c_k	Unit capacity cost of the supplier	Exogenous, deterministic, $w > c_p + c_k$
p	Unit product price	Exogenous, deterministic, $p > w$
α	Proportional distress cost stated as a portion $(1 - \alpha)$ of raised capital upon reorganization	Exogenous, deterministic, $\alpha \in (0, 1)$
β	Portion of orders at wholesale price in dual-financing scheme	Decision variable, $\beta \in [0, 1]$
γ	Proportional liquidation cost stated as a portion $(1 - \gamma)$ of firm value upon financial default	Exogenous, deterministic, $\gamma \in (0, 1)$
δ	Portion of the financial institution's loss compensated by the retailer upon supplier liquidation	Exogenous, $\delta \in (0, 1)$
λ	Portion of purchase order value received by the supplier from the financial institution	Decision variable, $\lambda \in [0, \bar{\lambda}], \bar{\lambda} \in (0, 1)$

(continued)

Table 7.1 (continued)

Symbol	Description	Assumptions
r	Interest rate of BPOF loan offered by the financial institution to the supplier	$r \geq 0$
r_f	Risk-free interest rate	$r_f = 0$
r_s	Commercial bank interest rate for the supplier	$r_s = \infty$

*Throughout the chapter we use "s", "r", and "c" (for creditor) to denote the "supplier", "retailer", and "financial institution", respectively

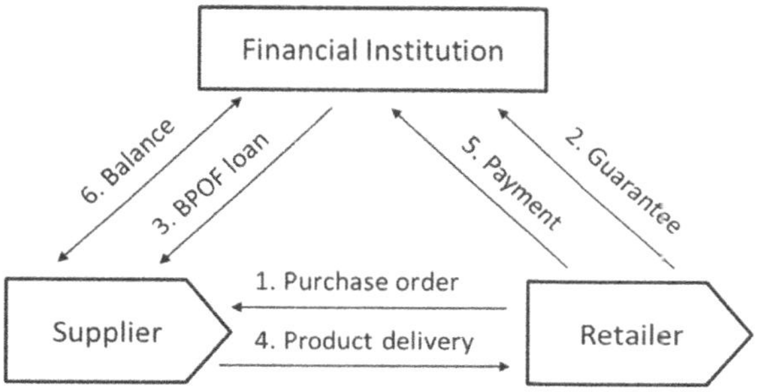

Fig. 7.2 Sequence of events in buyer-backed purchase order financing

risk-free interest rate r_f is normalized to zero (Brennan et al. 1988; Jing et al. 2012); this allows us to concentrate on the *effective* rate of APD and BPOF (i.e., the actual rate above risk-free rate).

Buyer-backed purchase order financing is the only pre-shipment option for external financing. In this case, the retailer initiates a tripartite agreement with a financial institution to provide the supplier with a BPOF loan based on the retailer's guarantee. Figure 7.2 depicts the sequence of events in BPOF. In this setting, other types of external financing are not accessible to the SME supplier because of his low credit rating.

Hence, the loan interest rate offered by commercial banks to the supplier is assumed w.l.o.g. as $r_s = \infty$. We assume that a backup supplier is not available. In practice, this captures two cases: (i) when there is no alternative supplier; and (ii) when the lead time needed to qualify a backup supplier is prohibitively long. For instance, BMW sources the sun roofs for its Z4 convertible from the supplier Edscha, which filed for bankruptcy in February 2009. Even if BMW could find an alternative supplier, at least six months would elapse before the new company could start producing the convertible top (Sodhi and Tang 2012). These considerations lead the buyer to seek pre-shipment finance instruments to fund the supplier's working capital.

Since neither APD nor BPOF can be realized without the retailer's agreement, she will be able to choose between these two financing strategies. In order to focus on financing schemes, our model is based on a wholesale price only contract owing to its simplicity and wide use in practice. In order to avoid the cost of financial distress and

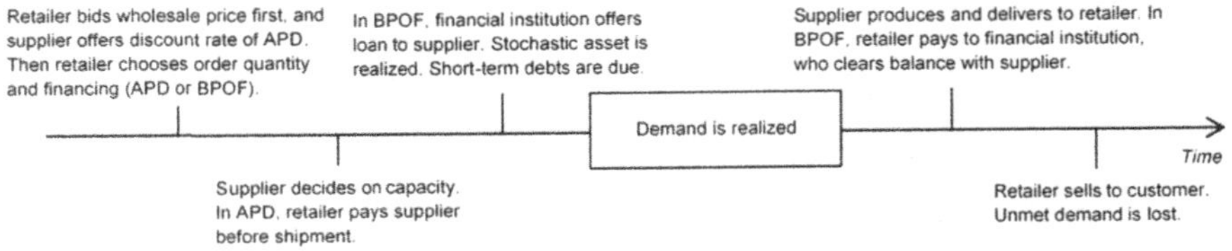

Fig. 7.3 Timeline of events

bankruptcy, the SME supplier may opt to install less capacity. Thus, the retailer is incentivized to assist the supplier with financing and thereby ensure order delivery. Capacity building is associated with lead time, and there is no lead time for production. Before the sales season, the supplier installs his capacity $K \in \mathbb{R}_+$ at unit capacity cost c_k. Then he manufactures at unit production cost c_p during the sales season. The salvage value of capacity is zero.

The timeline of events is presented in Fig. 7.3. First, the retailer offers wholesale price w to the supplier. If the supplier accepts the wholesale price, he will offer the unit discount rate d. The supplier ensures $w(1-d) > c_p + c_k$ for his profitability. Then the retailer decides on an order quantity q and chooses the financing strategy (i.e., APD or BPOF). The supplier determines his capacity K based not only on the retailer's order quantity but also on the extent to which his working capital is financed. In order to discourage underinvestment by a financially distressed supplier, the retailer is incentivized to provide pre-shipment finance through APD or BPOF. If the retailer chooses APD, she pays the supplier before product shipment. If the retailer chooses BPOF, she provides a guarantee such that a financial institution offers a BPOF contract $(\bar{\lambda}, r)$ to the supplier based on the retailer's credit rating. Here $\bar{\lambda} \in (0, 1)$ is the upper bound on the share of the purchase order that the financial institution is willing to lend, and r is the BPOF interest rate. If the supplier accepts the BPOF contract then he set (to the extent allowed by that contract) a borrowing level $\lambda \in [0, \bar{\lambda}]$, after which he receives BPOF in the amount of λwq from the financial institution. Then the supplier's stochastic internal asset level is realized. Short-term debts of both the supplier and retailer are due before demand realization.

The sales season begins and demand is realized, then the supplier produces and delivers to the retailer. In the base case, the retailer pays $\min(q, K)$ to the supplier. In APD, the retailer would receive a refund from the supplier if any prepaid order quantity is not fulfilled. In BPOF, if the supplier operates in continuation or undergoes reorganization then the retailer pays the total purchase order amount to the financial institution, which deducts the BPOF loan's principal and interest and then pays the supplier the balance. If the supplier is in liquidation, then (i) the financial institution seizes the supplier's liquid assets and (ii) the retailer pays a previously negotiated portion δ of the financial institution's loss as compensation, because the BPOF loan was backed by her guarantee. The retailer sells products to customers at a unit price p that is determined by the market. We assume $p > w$ to ensure that the retailer is profitable. Unmet demand is lost, and the salvage value of unsold product is zero.

7.2.1 Centralized and Decentralized Benchmarks

To establish benchmarks for a capital-constrained supply chain, we analyze the cases of centralized and decentralized supply chains that do not exhibit financial distress. In a *centralized* supply chain, an integrated firm with ample internal capital makes capacity decisions geared to achieving optimal channel-wide expected profit. The expected profit of an integrated firm that has made capacity decision K is[4]

$$\pi^{\mathrm{csc}}(K) = (p - c_p)\mathbb{E}\min(K, D) - c_k K. \tag{7.1}$$

The profit function is concave, and the optimal solution is $K^{\mathrm{csc}} = F^{-1}(\frac{p-c_p-c_k}{p-c_p})$.

In a *decentralized* supply chain, both the retailer and supplier have sufficient internal capital. The retailer's problem is equivalent to (1) except that she orders inventory at wholesale price w instead of producing it at cost c_p and c_k. Hence her expected profit is

$$\pi_r^{\mathrm{dsc}}(q, w) = p\mathbb{E}\min(D, \min(q, K)) - w\min(q, K). \tag{7.2}$$

The profit function is concave, and its optimal solution is $q^{\mathrm{dsc}} = F^{-1}\left(\frac{p-w}{p}\right)$. In contrast, for a centralized supply chain the optimal solution is $q^{\mathrm{dsc}} = F^{-1}\left(\frac{p-w}{p}\right) < K^{\mathrm{csc}} = F^{-1}(\frac{p-c_p-c_k}{p-c_p})$; this expression reflects the result of "double marginalization" (Spengler 1950).

The supplier anticipates the retailer's order for any wholesale price. He faces demand curve $q^{\mathrm{dsc}}(w)$ and chooses optimal capacity K^{dsc} to maximize his profit:

$$\pi_s^{\mathrm{dsc}}(K, w) = (w - c_p)\min(q, K) - c_k K. \tag{7.3}$$

We follow the approach of Lariviere and Porteus (2001, p. 295) and write the supplier's inverse demand curve as $w(q) = p\bar{F}(q)$. We define $\hat{q}$ as the least upper bound on the set of points such that $v(q) \geq 1$. Here the price elasticity of the retailer's order is $v(q) = -w(q)/[qdw(q)/dq]$ if we assume that $v'(q) \leq 0$ for $q \in [a, b)$ and $\hat{q} \in [a, \infty)$.

Lemma 7.1 *In a decentralized supply chain with sufficient capital, the supplier's first-order condition is*

$$p\bar{F}\left(K^{\mathrm{dsc}^*}\right)\left[1 - h\left(K^{\mathrm{dsc}^*}\right)\right] = c_k + c_p. \tag{7.4}$$

---

[4]We shall use "csc" and "dsc" to denote (respectively) the "centralized supply chain" and "decentralized supply chain" benchmarks. For brevity we use q, w, and K to denote (respectively) q^i, w^i, and K^i for $i =$ csc, dsc, bc, bpof, apd, df, where "bc" denotes "base case" and "df" denotes "dual financing".

The supplier's profit is unimodal on $[0,\infty)$, linear and strictly increasing on $[0,a)$, strictly concave on $[a,\hat{q}]$, and strictly decreasing on $(\hat{q},\infty)$. The optimal solution q^{dsc} to Eq. (7.4) is unique and must reside on the interval $[a,\hat{q}]$. The supplier's optimal capacity is $K^{\mathrm{dsc}}=q^{\mathrm{dsc}}$, his optimal sales quantity is either q^{dsc} or a, and the optimal wholesale price is $w^{\mathrm{dsc}}=p\bar{F}\left(q^{\mathrm{dsc}}\right)=\frac{c_k+c_p}{1-h(q^{\mathrm{dsc}})}$.

Proof. All proofs are given in the Appendix (available on request).

7.2.2 Base Case

In this case, one established retailer orders from one SME supplier, where each party could be capital constrained and neither APD nor BPOF is viable. Hence, the retailer's expected profit is $\pi_r^{\mathrm{bc}}(q,w)=p\mathbb{E}\min(D,\min(q,K))-w^{\mathrm{bc}}\min(q,K)$. Suppose the capital-constrained supplier's capacity decision is K^{bc}; then his profit function is

$$\pi_s^{\mathrm{bc}}(K,w)=\begin{cases}\left(w-c_p\right)\min(q,K)-c_kK & \text{in continuation,}\\ \left(w-c_p\right)\min(q,K)-c_kK-(1-\alpha)(L_s-A_s+c_kK) & \text{in reorganization,}\\ 0 & \text{in liquidation.}\end{cases}$$

If the supplier's liquid asset is able to cover current obligations (i.e., if $A_s\geq L_s+c_kK$) before order delivery, then the supplier continues his operations.[5] Otherwise, the supplier chooses between two alternatives to bankruptcy, reorganization or liquidation; he chooses the latter only if the cost of financial distress exceeds the firm's operating profit under reorganization (Yang et al. 2015). That is, if the supplier expected profit after reorganization is positive, then he will opt to undergo a costly reorganization process. Therefore, the supplier's probability of continuation is $\Pr(c)=\bar{\Phi}(\widehat{A_s})$ for $\widehat{A_s}\triangleq L_s+c_kK$, the probability of liquidation is $\Pr(l)=\Phi(\widetilde{A_s})$ for $\widetilde{A_s}\triangleq L_s+c_kK-\frac{(w-c_p)\min(q,K)-c_kK}{(1-\alpha)}$, and the probability of reorganization is $\Pr(r)=\Phi(\widehat{A_s})-\Phi(\widetilde{A_s})$. In order to avoid the costs associated with financial distress or default, a supplier may install less capacity (i.e. under-investment) than the retailer's order quantity.

In case of liquidation, the supplier's profit is zero thus his optimal capacity $K^{\mathrm{bc}^*}=0$. In the scenarios of continuation and reorganization, the supplier's optimal capacity decisions K^{bc} can be derived from the first-order conditions (FOCs):

[5]The conditions for continuation, reorganization, and liquidation can be readily derived, throughout the chapter, using similar logic. So to ease the exposition, hereafter we omit the corresponding analytical expressions.

$$\begin{aligned} p\bar{F}\left(K^{\mathrm{bc}^*}\right)\left[1-h\left(K^{\mathrm{bc}^*}\right)\right] &= c_p + c_k && \text{in continuation,} \\ p\bar{F}\left(K^{\mathrm{bc}^*}\right)\left[1-h\left(K^{\mathrm{bc}^*}\right)\right] &= c_p + (2-\alpha)c_k && \text{in reorganization} \end{aligned} \tag{7.5}$$

The supplier's optimal capacity can be derived by taking the weighted average of the optimal capacity in each scenario. Here $w^{\mathrm{bc}^*} = \frac{(c_p + c_k)\Pr(c) + (c_p + (2-\alpha\, c_k)\Pr(r)}{1-h(K)}$. The FOCs indicate that the supplier trades off the marginal benefit of unit production against the marginal cost of unit capacity.

7.3 Financing with APD or BPOF

To establish the basis for analyzing the financing equilibria of APD and BPOF, we first examine the cases where only one of the two instruments is viable in the focal supply chain.

7.3.1 Buyer-Backed Purchase Order Financing

We start with the case in which the retailer adopts only buyer-backed purchase order financing. Hence, she contacts a financial institution to establish a line of credit for the supplier based on her purchase order (see Fig. 7.2).

Recall from Sect. 2.1 that the retailer initiates BPOF by providing a loan guarantee (on the supplier's behalf) to the financial institution, which then offers a BPOF contract $(\bar{\lambda}, r)$ to the supplier. A supplier who accepts a BPOF contract can choose a borrowing level $\lambda \in [0, \bar{\lambda}]$, whereafter he receives a BPOF loan λwq from the financial institution. If the supplier repays his short-term debt L_s prior to product shipment, then he continues to operate in the sales season; otherwise, the supplier files for bankruptcy and chooses between reorganization and liquidation. The supplier liquidates when his internal asset falls below the threshold value $\widetilde{A_s} = L_s + c_k K - \lambda wq - \frac{(w-c_p)\min(q,K) - c_k K - \lambda wqr}{1-\alpha}$, thus the probability of liquidation is $\Pr(l) = \Phi\left(\widetilde{A_s}\right)$. The supplier ships his products after demand realization. In the event of continuation (continued operation) or reorganization, the retailer pays back the financial institution, which deducts the BPOF loan principal and interest before remitting the balance to the supplier. In the event of liquidation, however, the POF loan is in default; then the financial institution receives the supplier's liquid assets and the retailer—as guarantor of the supplier's credit—pays a previously determined portion (δ) of the financial institution's loss. Any additional fees charged by the financial institution are normalized to zero because they are usually small and do not affect our structural results.

We use backward induction to derive optimal decisions in the supply chain. In BPOF-only scenario, the retailer's expected profit is

$$\pi_r^{\text{bpof}}(q,w) = p\mathbb{E}\min(D,\min(q,K)) - w\min(q,K) - \int_{\underline{A_s}}^{\widetilde{A_s}} \delta\{\lambda wq - \gamma[(w-c_p)\min(q,K) - c_k K - \lambda wqr + A_s - L_s]\}\phi(A_s)dA_s. \tag{7.6}$$

The retailer will agree to guarantee the supplier's credit if her expected profit from initiating BPOF exceeds that in the base case—that is, if $\pi_r^{\text{bpof}}(q,D,K) \geq \pi_r^{\text{bc}}(q,D,K)$. Hence the likelihood of buyer participation in BPOF declines as the supplier's liquidation probability $\Phi\left[\widetilde{A_s}\right]$ increases.

The supplier makes operational and financial decisions to maximize his profit:

$$\pi_s^{\text{bpof}}(K,w,\lambda) = \begin{cases} (w-c_p)\min(q,K) - c_k K - \lambda wqr & \text{in continuation,} \\ (w-c_p)\min(q,K) - c_k K & \text{in reorganization,} \\ \quad -(1-\alpha)(L_s - A_s - \lambda wq + c_k K) - \lambda wqr & \\ 0 & \text{in liquidation.} \end{cases}$$

We denote the supplier's break-even BPOF interest rate as $\hat{r}$, and his optimal borrowing level (λ^*) is defined in Proposition 7.1.

Proposition 7.1 *In the case of buyer-backed purchase order financing only, the supplier's optimal borrowing level is*

$$\lambda^* = \begin{cases} \arg\max\limits_{\lambda\in(0,\bar{\lambda}]} \pi_s(\lambda,w,K) & \text{if } r<1-\alpha \text{ and } r\leq\hat{r}, \\ 0 & \text{otherwise;} \end{cases}$$

In case of liquidation, his optimal capacity $K^{\text{bpof}^*} = 0$. *Otherwise* K^{bpof^*} *satisfies the FOCs*

$$\begin{aligned} p\bar{F}\left(K^{\text{bpof}^*}\right)\left[1-h\left(K^{\text{bpof}^*}\right)\right] &= c_p + c_k && \text{in continuation,} \\ p\bar{F}\left(K^{\text{bpof}^*}\right)\left[1-h\left(K^{\text{bpof}^*}\right)\right] &= c_p + (2-\alpha)c_k && \text{in reorganization.} \end{aligned} \tag{7.7}$$

Proposition 7.1 indicates that if capital constraints lead the supplier to underinvest in capacity, then the risk of that scenario's transpiring can be mitigated by BPOF at the cost of an interest payment. The condition for a positive borrowing level, i.e., $r<1-\alpha$ and $r\leq\hat{r}$ ensures the financing cost of BPOF is lower than proportional distress cost and is profitable for the supplier. If the BPOF loan is sufficient to fund the supplier's working capital ($\bar{\lambda} \geq \frac{c_k K^{\text{bpof}} + L_s - \underline{A_s}}{wq^{\text{bpof}}}$) and $r\leq\hat{r}$, then the supplier invests in capacity $K^{\text{bpof}} = q^{\text{bpof}} = F^{-1}\left(\frac{p-w}{p}\right)$, i.e., as if he were not capital constrained. In other words, the supplier can make his operational decisions without considering financial constraints *provided* the BPOF loan is sufficient for his working capital needs.

Anticipating the supplier's response, the financial institution decides on $(\bar{\lambda}, r)$ to maximize its expected profit:

$$\begin{aligned}\pi_c =& \bar{\Phi}(\widetilde{A_s})\lambda wqr \\ &+ \int_{\underline{A_s}}^{\widetilde{A_s}} (1-\delta)\{\gamma[(w-c_p)\min(q,K) - c_k K - \lambda wqr + A_s - L_s] - \lambda wq\}\phi(A_s)dA_s.\end{aligned} \tag{7.8}$$

The first term on the right-hand side is the financial institution's expected payoff when the supplier is in continuation or reorganization, and the second term is its expected payoff when the supplier defaults and liquidates. Although the exact number of BPOF providers is not easy to estimate, the *New York Times* reports that there are at least six major BPOF companies in the US market (Martin 2010). We therefore assume that the BPOF lending market is competitive, from which it follows that the financial institution has a zero expected payoff:

$$\begin{aligned}&\lambda wqr\left[\bar{\Phi}\left(\widetilde{A_s}\right) - (1-\delta)\Phi\left(\widetilde{A_s}\right)\gamma\right] + \int_{\underline{A_s}}^{\widetilde{A_s}} (1-\delta) \\ &\quad\gamma[(w-c_p)\min(q,K) - c_k K + A_s - L_s]\phi(A_s)dA_s = (1-\delta)\Phi\left(\widetilde{A_s}\right)\lambda wq.\end{aligned} \tag{7.9}$$

.

Equation (7.9) reveals that r increases with λ when $\bar{\Phi}\left(\widetilde{A_s}\right) > (1-\delta)\Phi\left(\widetilde{A_s}\right)\gamma$. In other words, if the joint probability of continuation and reorganization is *greater* than the liquidation probability multiplied by both the bankruptcy cost ratio and the loss compensation factor—a condition frequently satisfied in practice—then the financial institution will charge a higher interest rate if the supplier's expected borrowing level is higher. The underlying reason is that lending risk increases with loan size.

Buyer-backed purchase order financing enables the supplier to receive working capital funding from a financial institution based on the retailer's credit rating, which creates a "win–win–win" situation by providing three benefits: (i) mitigating the retailer's supply shortfall due to the supplier's financial distress, (ii) financing the supplier's working capital to ensure that orders are fulfilled (at the cost of BPOF interest), and (iii) yielding the financial institution the BPOF interest payment. In comparison with advance payment discount that transfers capital within the supply chain, using a financial institution's capital via BPOF enables the supplier to fulfill order requirements and the retailer to extend payment terms.

Our observations are in line with empirical evidence that BPOF benefits not only supplier and buyer but also the financial institution (Klapper 2006; Navas-Alemán et al. 2012). For the supplier, BPOF reduces transaction and borrowing costs. Buyer-backed purchase order financing offers working capital at favorable rates, which provides instant liquidity by reducing days sales outstanding (DSO) and thereby accelerates the cash-to-cash (C2C) cycle (Farris and Hutchison 2002). From the buyer's perspective, since her accounts payables are managed by a financial institution, her administrative costs of processing different payment terms with multiple suppliers are reduced. By financing the suppliers' working capital, the buyer can enhance her reputation and relationships with SME suppliers. For the financial institution, BPOF helps to develop relationships with suppliers. For example, credit histories of SME suppliers can be assembled based on BPOF. Since BPOF involves only high-quality receivables, financial institutions can expand their operations without increasing credit risk.

7.3.2 Advance Payment Discount

In this case, the retailer adopts only advance payment discount. Here we assume that the retailer purchases the *total* order quantity at a discounted price before order delivery (this assumption is generalized in Sect. 4.2). Although the retailer is not at risk of bankruptcy, she may be subject to short-term financial distress if she finances the supplier solely via advance payment discount. The retailer offers wholesale price w and the supplier offers advance payment discount rate $d \in \left[0, \frac{w-c_p-c_k}{w}\right]$. Then the retailer decides on order quantity q, adopts APD only by paying $wq^{\mathrm{apd}}(1-d)$ to the supplier before product shipment, and will receive $w(1-d)\left(q^{\mathrm{apd}}-K^{\mathrm{apd}}\right)^+$ after delivery in case any amount ordered is not fulfilled.

In this APD-only scenario, the supplier receives total order payment before product delivery. When APD is sufficient to fund the supplier's working capital $(w(1-d)q \geq c_k K^* + L_s - \underline{A_s})$, he can set his optimal capacity $K^{\mathrm{apd}^*} = q^{\mathrm{apd}^*}$ without experiencing financial distress, and his profit is $\pi_s^{\mathrm{apd}}(K, w) = \left[w(1-d) - c_p\right]\min(q, K) - c_k K$. In contrast, the retailer with initial asset A_r and short-term liability L_r (due before order delivery) might incur financial distress. Given her internal asset, the retailer chooses APD as long as it does not lead to bankruptcy with liquidation (otherwise, the retailer will not adopt APD); her aim is to maximize the expected profit $\pi_r^{\mathrm{apd}}(K, w) = p\mathbb{E}\min[D, \min(q, K)] - w(1-d)\min(q, K) - (1-\alpha)\left[L_r + w(1-d)q - A_r\right]^+$.

Proposition 7.2 *In case that advance payment discount is sufficient to fund the supplier's working capital, the retailer's optimal order quantity q^{apd} satisfies*

$$
\begin{aligned}
p\bar{F}\left(q^{\mathrm{apd}}\right) &= w(1-d) && \textit{in continuation,} \\
p\bar{F}\left(q^{\mathrm{apd}}\right) &= w(1-d)(2-\alpha) && \textit{in reorganization.}
\end{aligned}
\tag{7.10}
$$

Moreover, $q^{\mathrm{apd}} \geq q^{\mathrm{dsc}}$ if and only if $(1-d)(2-\alpha) \leq 1$.

This proposition demonstrates the retailer's trade-off between marginal effect of advance payment discount and unit cost of financial distress when the retailer's internal asset level is below a certain threshold ($A_r < L_r + w(1-d)q^{\text{apd}}$). When the *effective* wholesale price (i.e., the price that determines the retailer's order quantity) $w_e \triangleq w(1-d)[1+(1-\alpha)1_{L_r + w(1-d)q^{\text{apd}} > A_r}]$ does not exceed w, the channel coordination benefit of APD dominates the financial distress effect; hence $q^{\text{apd}} \geq q^{\text{dsc}}$. In one special case where a well-established retailer endowed with internal capital reserves and thus immune to financial distress, $w_e \leq w$. Thus the APD's channel coordination effect is guaranteed: $q^{\text{apd}} \geq q^{\text{dsc}}$. If the effective wholesale price $w_e > w$, then the channel coordination benefit of APD is dominated by its financial distress effect and thus $q^{\text{apd}} < q^{\text{dsc}}$.

The retailer benefits from APD's channel coordination effect but at the cost of potential financial distress. The supplier also benefits from that effect as well as from the faster collection of receivables (Farris and Hutchison 2002). Both APD and BPOF can relieve the supplier's capital constraint: BPOF increases the supply chain's overall financing capacity by borrowing capital from a financial institution, whereas APD transfers monetary flows from voluntary to binding positions within the supply chain. In addition to mitigating the supplier's budget constraint, advance payment discount reduces the wholesale price and thus yields a coordination benefit for the supply chain (though at the cost of the retailer's possible financial distress).

7.4 Interactions Between APD and BPOF

So far we have derived the respective optimal decisions in supply chain under APD and under BPOF. If both APD and BPOF are viable then one question naturally arises: which trade finance instrument(s) will be selected in the financing equilibrium? To answer this question, Sect. 4.1 characterizes the retailer's optimal choice between APD and BPOF when only one of the two instruments can be adopted. In Sect. 4.2, we present the optimal financing and operational decisions when both APD and BPOF can be chosen simultaneously.

7.4.1 *Single Financing Equilibrium Between APD and BPOF*

When either APD or BPOF alone is sufficient to fund the supplier's need for working capital (i.e., when $wq^*\bar{\lambda} \geq c_k K^* + L_s - \underline{A_s}$ and $w(1-d)q \geq c_k K^* + L_s - \underline{A_s}$), the retailer will choose one of the instruments in *single-financing* scheme. We assume w.l.o.g. that the retailer chooses APD over BPOF if she is indifferent between them.

Theorem 7.1 *Under single financing, there exists a unique threshold of the retailer's internal asset level ω_r such that she prefers BPOF if and only if $A_r < \omega_r$. Otherwise, the retailer chooses APD.*

This threshold of the retailer's internal asset implies that the optimal financing depends: (i) under APD, on the trade-off between the benefit of channel coordination and the cost of financial distress; and (ii) under BPOF, on the trade-off between the benefit of a secured supply and the cost of a credit guarantee in case the supplier liquidates. Although the retailer pays a lower price in APD than in BPOF, she then faces increased inventory risk (because the order quantity is greater) in addition to the risk of financial distress. When the retailer's internal asset level is so high that financial distress is not a concern, she always prefers APD to BPOF.

Theorem 7.1 indicates that a more wealthy buyer prefers using internal sources of capital to fund the supplier whereas a less wealthy buyer—whose credit rating is higher than her supplier's—should leverage that interest rate spread to borrow capital from a financial institution. The implication is that APD is most suitable for a supply chain consisting of a capital-constrained supplier and an established retailer whose high internal asset level precludes financial distress; this follows because the retailer prefers APD to BPOF provided the former's channel coordination benefits dominate her concerns about financial distress. These observations conform with the empirical finding that APD accounts for 19–22% of global trade finance (Chauffour and Malouche 2011), since established and wealthier retailers with larger trade volumes are more likely to adopt APD. Because BPOF transfers credit risk from a distressed supplier to a high-quality buyer, it is frequently used for transactions between a buyer in a developed economy and an SME supplier located in a developing economy (Gold and Jacobs 2007; USAID 2011a, b).

7.4.2 *Dual Financing with Both APD and BPOF*

If either APD or BPOF alone is not sufficient to cover the supplier's working capital needs, then both instruments can be adopted in a *dual-financing* scheme. Under dual financing, the retailer orders $q^{\text{df}} = q_1 + q_2$ from the supplier. The terms $q_1 = (1-\beta)q^{\text{df}}$ and $q_2 = \beta q^{\text{df}}$ ($\beta \in [0, 1]$) denote the respective order quantities paid at advance discount before shipment and at wholesale price after delivery. The retailer maximizes her profit by jointly choosing the total order quantity q^{df} and the portion of order quantity paid at wholesale price β. Dual financing is viable when the following two conditions hold: (i) the BPOF loan is insufficient to cover the supplier's working capital (i.e., $wq_1^*\bar{\lambda} < c_k K^* + L_s - \underline{A_s}$; this generalizes the assumption in Sect. 4.1); and (ii) the retailer's *partial* payment via APD is insufficient to cover the supplier's working capital needs (i.e., $w(1-d)q_1^* < c_k K^* + L_s - \underline{A_s}$; this relaxes assumptions in Sects. 3.2 and 4.1). Therefore, dual financing represents the general case of pre-shipment finance.

In a dual-financing scheme, the retailer chooses (q^{df}, β) in order to maximize her expected profit: $\pi_r^{\text{df}}(q, w) = p\mathbb{E}\min(D, q^{\text{df}}) - wq_2 - w(1-d)q_1 - (1-\alpha)[L_r + w(1-d)q_1 - A_r]^+$.

Theorem 7.2 *If dual financing is sufficient to fund the supplier's working capital, the retailer's optimal order quantity satisfies the FOCs*

$$\begin{aligned} pF\left(q^{\mathrm{df}}\right) &= p - w(1-d) - \beta wd && \textit{in continuation,} \\ pF\left(q^{\mathrm{df}}\right) &= p - (2-\alpha-\beta+\alpha\beta)w(1-d) - \beta wd && \textit{in reorganization.} \end{aligned} \tag{7.11}$$

The supplier's optimal BPOF borrowing level is

$$\lambda^* = \begin{cases} \arg\max\limits_{\lambda\in(0,\bar{\lambda}]} \pi_s(\lambda, w, K) & \textit{if } r<1-\alpha \textit{ and } r\leq\hat{r}, \\ 0 & \textit{otherwise.} \end{cases}$$

Here $\hat{r}$ denotes the supplier's break-even BPOF interest rate under dual financing. Theorem 7.2 shows that, in APD, the retailer trades off the benefit of a unit discount against the possible cost of financial distress. If APD proves insufficient, the supplier then seeks external financing via BPOF to cover working capital needs. In dual financing, a retailer with sufficient internal capital favors funding of the supplier by APD whereas a retailer with insufficient capital will first maximize her use of APD and only then use BPOF when the cost of financial distress outweighs the benefit of a price discount.

Lemma 7.2 *In the case of continuation, the retailer's total order quantity is decreasing in the proportion β of order quantity paid at wholesale price after product shipment. In the case of reorganization, the retailer's total order quantity is increasing in β if and only if $(1-\alpha) > \frac{d}{1-d}$.*

Lemma 7.2 implies that the retailer strictly prefers APD to BPOF if her internal capital is high enough to preclude financial distress. If the retailer goes through reorganization, she may then prefer BPOF to APD under the binding condition that the marginal cost of financial distress is greater than the benefit of a unit discount.

Theorem 7.2 and Lemma 7.2 indicate that the choice of an optimal pre-shipment finance strategy depends on the retailer's level of internal capital. As a risk-sharing scheme that provides channel coordination benefits despite (possibly) incurring financial distress, APD receives priority under dual financing when the retailer possesses ample internal capital.

7.5 Numerical Examples

In this section, we examine the robustness of our main results via an extensive numerical study. Examples are provided to demonstrate how demand variability and the retailer's internal capital level each affects supply chain performance. Our numerical study relies on optimization via simulation using Palisade @RISK software with 1,000,000 uncertainty scenarios (supply–demand matching states and price–quantity decisions) and the following benchmark parameter values: $p = 60$,

$c_k = c_p = 10$, $\alpha = 0.85$, and $\gamma = 0.9$ (cf. Gamba and Triantis 2014). Demand D follows a normal distribution, $N(1000, 100)$, whose standard deviation may vary with its coefficient of variation (CV).

7.5.1 *Impact of Demand Variability*

We first examine how demand volatility affects the interaction between APD and BPOF. When adjusting demand volatility, we allow its coefficient of variation $\delta(D)$ to vary between 0.1 and 0.5; at the same time, we use a constant value for expected demand, $\mathbb{E}[D] = 1000$. We compare the channel efficiency under APD and BPOF and plot the threshold points in Figs. 7.4 and 7.5. For each financing scheme, we follow Jing et al. (2012) in examining the percentage competition penalty (i.e., the portion of profit decrease in decentralized supply chain relative to centralized supply chain)

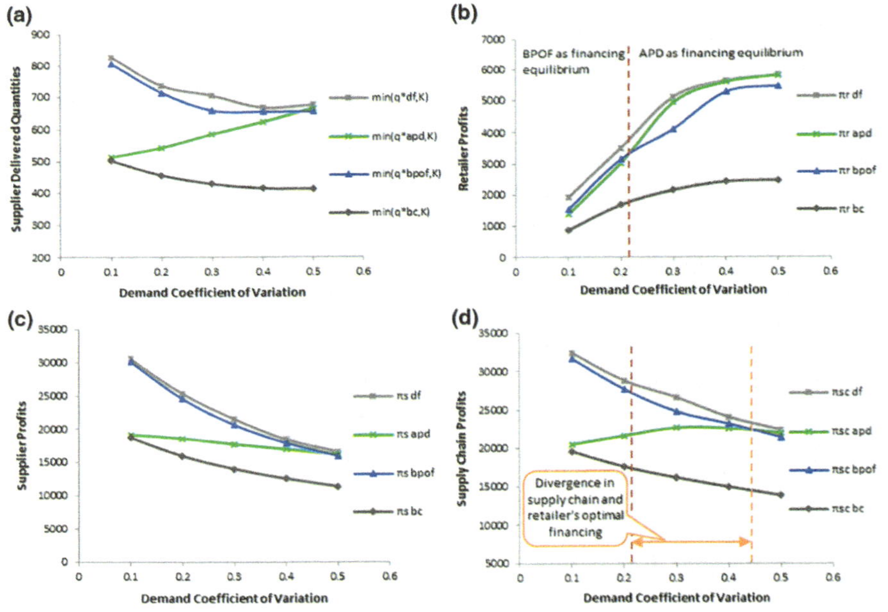

Fig. 7.4 Effect of demand volatility on **a** supplier's delivered quantities; **b** retailer's profits; **c** supplier's profits; and **d** supply chain profits

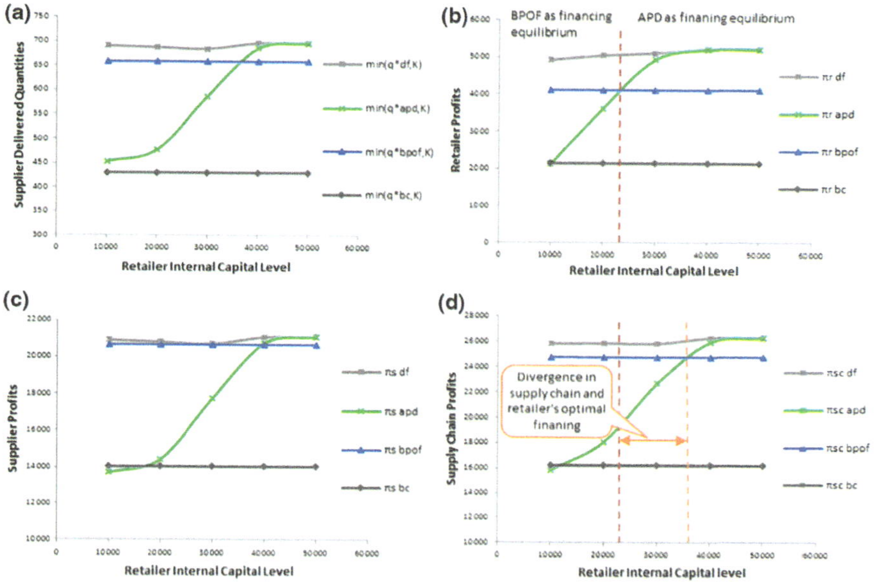

Fig. 7.5 Effect of retailer's internal capital level on **a** supplier's delivered quantities; **b** retailer's profits; **c** supplier's profits; and **d** supply chain profits

$$\mathcal{P} \triangleq \left(1 - \frac{\pi_s + \pi_r}{\pi^{\text{csc}}}\right) \times 100\%$$

w.r.t. demand volatility and the retailer's internal capital level. Figure 7.4 presents the effects of increasing demand CV on supply chain performance and on the financing equilibrium between APD and BPOF.

Figure 7.4a shows that the supplier's delivered quantity is decreasing in demand volatility without pre-shipment finance (APD or BPOF). Whereas the supplier's delivered quantity is strictly increasing in APD only (green curve), it is convex in BPOF only (blue curve) as well as in the dual-financing scheme (grey curve). Absent pre-shipment finance, the supplier's order delivery decreases as demand volatility increases. When pre-shipment finance (APD and/or BPOF) is employed, the supplier's delivered quantity is less sensitive to demand volatility because his financial distress is mitigated by pre-shipment finance; thus, demand risk is shared with the retailer. Figure 7.4b demonstrates that the retailer's profits are increasing in demand volatility as a result of the lower wholesale price. Comparing the retailer's profits under APD only with BPOF only reveals that the latter's equilibrium region decreases with demand variability. That is, in single financing, the retailer chooses BPOF when the demand coefficient of variation is relatively low (i.e., when $\delta(D) < 0.22$); otherwise, she chooses APD in equilibrium. When demand CV is sufficiently high, APD dominates BPOF and becomes the only strategy adopted in dual financing; in this case, APD only and dual financing yield the same payoff. It is intuitive that the retailer would prefer a wholesale price with BPOF over APD as means of reducing risk when demand variability rises, because a lower order quantity and late payment with external financing are both associated with less inventory and credit risk. However, our numerical example shows that the established retailer can opt to take more risk via APD—and thus generate a higher return —when demand risk increases.

Figure 7.4c illustrates that the supplier's profits decline with respect to increasing demand volatility. The decline is much less in APD-only scenario owing to risk sharing between supply chain members. The supplier's profits are reduced at the same decreasing rate in the BPOF-only and dual-financing cases (the latter follows given that, when BPOF can fully mitigate financial distress, it is the only mechanism adopted under dual financing). Contrary to Fig. 7.4b, the supplier prefers the APD-only to the BPOF-only scenario if the demand coefficient of variation is relatively high (i.e., $\delta(D) \geq 0.48$). Figure 7.4d indicates the supply chain's profit is concave under APD only owing to the channel coordination effect, while is decreasing w.r.t. demand volatility in the other three scenarios. In contrast to Fig. 7.4b, BPOF dominates APD unless the demand CV is fairly high (i.e., $\delta(D) \geq 0.44$) from the supply chain's perspective. Therefore, supply chain members incur a competition penalty when demand variability falls within a specific interval, $\delta(D) \in [0.22, 0.44)$. In this interval, the retailer's optimal financing decision deviates from the supply chain's optimal choice. Thus there is performance inefficiency due to supplier–retailer competition in a decentralized setting. For example, when $\delta(D) = 0.4$, the competition penalty $\mathcal{P} = 2.4\%$; when $\delta(D) = 0.3$, $\mathcal{P} = 8.3\%$. When $\delta(D) = 0.22$, the competition penalty is at its maximum: $\mathcal{P} = 17.5\%$. When demand variability falls out of the given interval, supply chain members' incentives are aligned thus channel-wide optimization can be achieved using BPOF as external financing or APD as internal financing. These results suggest that a channel coordination mechanism is desirable in single financing equilibrium.

7.5.2 Impact of Retailer's Internal Capital Level

In order to examine the role played by the retailer's internal capital level (i.e., $A_r - L_r$), we vary the expected value of her asset level A_r from 40,000 to 80,000 while keeping her short-term liability L_r at a constant level of 30,000. Figure 7.5 shows that the retailer's internal capital level has no significant effect except in the APD-only scenario, because it is the only scenario where the retailer faces the possibility of financial distress. With regard to the retailer's internal capital level, the APD-only value is increasing while values of the other three strategies are relatively stable for supply chain partners.

Figure 7.5a demonstrates that the supplier's delivered quantity is increasing under APD only but is insensitive to the retailer's internal capital level in the other three scenarios. The underlying reason is that the retailer's growing internal capital relieves financial distress and thus allows APD-based exploitation of supply chain coordination; in contrast, the retailer may suffer financial distress if she employs APD only. Similarly, Fig. 7.5b illustrates that the retailer's profit is increasing in her internal capital level under APD only; however, the increase is less significant when the retailer's internal capital exceeds a certain level (i.e., $A_r - L_r > 31,231$) because financial distress is mitigated by increasing capital. A comparison of APD only and BPOF only (green and blue curves) verifies Theorem 7.1 in that BPOF dominates APD in single financing equilibrium if and only if the retailer's internal capital is below a certain threshold, i.e., $A_r - L_r < \omega_r - L_r = 23,218$. Hence, the BPOF-only equilibrium region is decreasing in the retailer's internal capital level. Conversely, if the retailer's internal capital level is relatively high then APD dominates BPOF and could become the only strategy adopted in the dual-financing scheme, which verifies Lemma 7.2. Thus the retailer's APD-only payoff approaches her dual-financing payoff with increasing levels of internal capital.

Figure 7.5c indicates that the supplier's profit is increasing in the retailer's internal capital level under APD-only as a result of the effect plotted in Fig. 7.5a. Contrary to the result in Fig. 7.5b, the supplier prefers APD to BPOF when the retailer's internal capital level is relatively high, i.e., when $A_r - L_r \geq 39,526$. Figure 7.5d presents the net effect of the retailer's internal capital level on supply chain profits: the channel coordination effect of APD is increasing in the retailer's internal capital level. In contrast to Fig. 7.5b, APD dominates BPOF when the retailer's internal capital level exceeds 36,541 from the supply chain's perspective. Therefore, channel members incur a competition penalty if the retailer's internal capital level is within a certain interval, i.e., when $23,218 \leq A_r - L_r < 36,541$. Within this interval, the retailer's optimal financing choice does not align with the supply chain's optimal strategy. For instance, when $A_r - L_r = 30,000$ the competition penalty is $\mathcal{P} = 8.3\%$, whereas $\mathcal{P} = 21.2\%$ (the maximum) when $A_r - L_r = 23,218$. This high cost associated with competition penalty makes the supply chain inefficient, thus channel coordination is crucial when pre-shipment finance instruments are employed.

The managerial insights from Figs. 7.4 and 7.5 are as follows. First, greater demand uncertainty increases the need for risk sharing within the supply chain, while higher level of the retailer's internal capital enables financing to supplier by advance payment discount. As demand risk and the retailer's internal capital level increases, transferring monetary flow from voluntary to binding parties in the supply chain becomes more desirable than raising capital from a financial institution. Second, both the interval and magnitude of the competition penalty imply that —even though pre-shipment finance instruments can mitigate the supplier's capital constraint—the incentive alignment between both channel members is indispensable for reducing the costs of competition penalty.

7.6 Pre-shipment Finance in Supply Chain

This chapter considers a capital-constrained supply chain with a SME supplier selling to an established retailer via a wholesale price contract. The retailer chooses between two pre-shipment finance instruments—advance payment discount and buyer-backed purchase order financing—to manage the supplier's financial distress. We characterize the optimal supply chain decisions in APD and BPOF respectively, and show that BPOF enables the supplier to finance his working capital via a loan from a financial institution based on the retailer's credit rating, while APD can bring channel coordination benefit at the cost of the retailer's potential financial distress. When either APD or BPOF can be chosen, the retailer prefers APD to BPOF if her internal asset level is above a certain threshold. In dual-financing when both APD and BPOF are adopted, the retailer chooses APD and will not initiate BPOF unless the marginal cost of financial distress outweighs the benefit of unit discount. We find that the equilibrium region of APD is increasing in both the retailer's internal capital level and demand variability. Competition penalty can lead to considerable costs when either the demand variability or the retailer's internal capital level are within certain (respective) intervals. Therefore, coordination and collaboration between supply chain partners is essential for mitigating competition penalty.

7.6.1 Managerial Implications

Buyer-backed purchase order financing creates a "win–win–win" situation by providing the supplier working capital funding from a financial institution based on the retailer's credit rating, in which the supplier alleviates his budget constraint and thereby ensures order fulfillment, the retailer strengthens her relationship with supplier and secures product quantity without incurring financial distress, while the financial institution earns interest payment with lower credit risk (cf. Klapper 2006; Navas-Alemán et al. 2012). In comparison, advance payment discount provides channel coordination benefit to both the supplier and retailer and accelerates the collection of the supplier's receivables thus enhances his cash conversion cycle at the potential cost of the retailer's financial distress. Although both APD and BPOF can mitigate the supplier's financial distress before product shipment, APD transfers monetary flows from voluntary to binding positions within the supply chain, while BPOF enables capital raising from a financial institution and thereby increases the supply chain's overall financing capacity.

Advance payment discount features a risk-sharing scheme that provides a lower wholesale price to the retailer yet also exposes her to higher inventory risk with larger order quantity and the potential cost of financial distress depending on her internal capital level. Hence, a retailer with ample capital that precludes financial distress strictly prefers APD, while a less wealthy retailer with higher credit rating than her supplier should leverage on the interest rate spread by initiating BPOF with a financial institution (cf. Gold and Jacobs 2007; Chauffour and Malouche 2011). Greater demand variability increases the need for risk sharing in the supply chain by APD, and higher level of the retailer's internal capital guarantees that the channel coordination benefit of APD dominates potential cost of financial distress. The interval and magnitude of the competition penalty highlights the value of incentive alignment between supply chain partners.

References

Babich, V. (2010). Independence of capacity ordering and financial subsidies to risky suppliers. *Manufacturing & Service Operations Management, 12,* 583–607.

Brennan, M., Maksimovic, V., & Zechner, J. (1988). Vendor financing. *Journal of Finance, 43,* 1127–1141.

Cai, G., Chen, X., & Xiao, Z. (2014). The roles of bank and trade credits: Theoretical analysis and empirical evidence. *Production and Operations Management, 23,* 583–593.

Chauffour, J. P., & Malouche, M. (2011). *Trade finance during the great trade collapse.* Washington, D.C.: World Bank.

Farris, M. T., II, & Hutchison, P. D. (2002). Cash-to-cash: The new supply chain management metric. *Journal of Physical Distribution & Logistics Management, 32,* 288–298.

Gamba, A., & Triantis, A. J. (2014). Corporate risk management: Integrating liquidity, hedging, and operating policies. *Management Science, 60,* 246–264.

Gold, M., & Jacobs, G. (2007). Analysis of the potential for development of SME buyer-backed purchase order financing products. United States Agency for International Development. URL: http://pdf.usaid.gov/pdf_docs/Pnadm257.pdf Retrieved on September 24, 2013.

Hortaçsu, A., Matvos, G., Shin, C., Syverson, C., & Venkataraman, S. (2011). Is an automaker's road to bankruptcy paved with customers' beliefs? *American Economic Review: Papers of Proceedings,* 101, 93–97.

Jing, B., Chen, X., & Cai, G. (2012). Equilibrium financing in a distribution channel with capital constraint. *Production and Operations Management, 21,* 1090–1101.

Klapper, L. (2006). The role of factoring for financing small and medium enterprises. *Journal of Banking Finance, 30,* 3111–3130.

Kouvelis, P., & Zhao, W. (2012). Financing the newsvendor: Supplier vs. bank, and the structure of optimal trade credit contracts. *Operations Research, 60,* 566–580.

Lariviere, M. A., & Porteus, E. L. (2001). Selling to the newsvendor: An analysis of price-only contracts. *Manufacturing & Service Operations Management, 3,* 293–305.

Leland, H. E. (1994). Corporate debt value, bond covenants, and optimal capital structure. *Journal of Finance, 49,* 1213–1252.

Marr, K. (2009). Auto suppliers seek help in wake of Chrysler, GM woes. *Washington Post.* May 14, 2009. URL: http://www.washingtonpost.com/wp-dyn/content/article/2009/05/13/AR2009051303382.html. Retrieved on October 10, 2013.

Martin, A. (2010). The places they go when banks say no. *The New York Times.* January 30, 2010. URL: http://www.nytimes.com/2010/01/31/business/smallbusiness/31order.html. Retrieved on December 7, 2013.

Navas-Alemán, L., Pietrobelli, C., & Kamiya, M. (2012). *Inter-firm linkages and finance in value chains.* Working paper, Inter-American Development Bank.

Nussel, P., & Sherefkin, R. (2008). Plastech bankruptcy could shut all of Chrysler. *Automotive News.* URL: http://www.autonews.com/apps/pbcs.dll/article?AID=/20080204/OEM/302049890. Retrieved on October 2, 2013.

PurchaseOrderFinancing.com. (2016). PO financing rates made affordable. *Informative Reactions,* July 16, 2016. URL: http://dustyannex298.wikidot.com/blog:7. Retrieved on November 24, 2016.

Sodhi, M., & Tang, C. (2012). Tactical approaches for mitigating supply chain risks: Financial and operational hedging. In M. Sodhi & C. Tang (Eds.), *Managing supply chain risk* (pp. 109–133). Berlin: Springer.

Spengler, J. (1950). Vertical integration and antitrust policy. *Journal of Political Economy, 58,* 347–352.

Tice, C. (2010). Can a purchase order loan keep your business growing? June 17, 2010. URL: http://www.entrepreneur.com/article/207058. Retrieved on December 8, 2013.

United States Agency for International Development [USAID]. (2011a). Case studies in buyer-backed purchase order financing. Crimson Capital Corp.

United States Agency for International Development [USAID]. (2011b). Launching buyer-backed purchase order financing in east Africa. Crimson Capital Corp.

Yang, S. A., Birge, J. R., & Parker, R. P. (2015). The supply chain effects of bankruptcy. *Management of Science, 61,* 2320–2338.

Chapter 8
Research Overview of Operations-Finance Interface

In this chapter, we take a qualitative and quantitative approach to reviewing conceptual research on—as well as analytical models and empirical assessments of—the interactions between operations and finance. First, we take a quantitative approach to bibliometric analysis (Tang and Musa 2011; Fahimnia et al. 2015). In particular, we conduct citation and PageRank analyses to identify both frequently cited and high-impact papers. Moreover, we analyze a data set of 258 peer-reviewed journal articles (excluding conference papers) on the OFI (i.e., operations–finance interface) from 1958 to 2016. In this context, we summarize the conceptual, analytical, and empirical research of operations-finance interface. In addition, we identify eight research streams in the paper citation network and sketch the landscape of operations–finance interactions based on leading articles and recent advances in each stream. Thus we statistically characterize the maturity and potential of these streams of research before proposing directions and approaches for future explorations in this field.

8.1 Operations-Finance Interface

Operations and finance are two key functions that jointly drive business success. Building capacity requires investment, and product sales generate revenue—factors that connect material and financial flows in cash conversion cycle (CCC). The operations–finance interface involves jointly optimizing material, monetary, and information flows when exposed to various, interrelated risks. The seminal theorem of Modigliani and Miller (1958) marked an important starting point for research on the operations–finance interface by proposing that operational and financial decisions can be separated when one assumes symmetric information and perfect capital markets. These assumptions seldom hold simultaneously in practice, so models that incorporate various market frictions have sparked growing attention in this interdisciplinary field (Birge et al. 2007; Babich and Kouvelis 2015). Research topics

L. Zhao and A. Huchzermeier, *Supply Chain Finance*, EURO Advanced Tutorials on Operational Research, https://doi.org/10.1007/978-3-319-76663-8_8

include operational hedging, financial flexibility, supply chain risk management, enterprise risk management, integrated risk management, and supply chain finance as well as the financial implications of supply chain strategies. Conceptual research, analytical modelling, and empirical assessments have been conducted to evaluate the interactions between operations and finance. Given the wide variety of topics and methodologies, it is challenging for scholars to develop a grasp of the research streams involving the operations–finance interface. Despite the high visibility of several often cited articles, the overall research landscape may remain ambiguous.

Our quantitative approach to bibliometric analysis is in contrast to extant literature reviews on the operations–finance interface, which are of a qualitative nature (as befits conceptual explorations and research summaries). Most previous syntheses of such research concentrate on analytical approaches; only a few review articles incorporate empirical assessments; and conceptual research has received limited attention. An overview of related papers and new contributions is given in Table 8.1.

In this chapter, we aim to explore three research questions:

(i) What are the recent publication trends in terms of article quantity on the operations–finance interface? (ii) What are the primary research streams and which papers have the greatest impact in each stream? (iii) Which research streams are already well explored, and which are relatively incipient and thus call for additional scholarly attention?

Our goals are reflected in this chapter's four main aspects. First, we adopt a quantitative approach to show the publication trends of conceptual, analytical, and empirical research on the OFI. Second, we identify the lead articles as measured by citation and PageRank analyses. Third, we classify the research into eight research streams to characterize the research landscape of this field. Fourth, we summarize the lead articles and recent publications in each stream to facilitate navigation of concepts, analytical models, and empirical evidence.

8.2 Research Methodology

This chapter aims to highlight emerging concepts and techniques, map out a holistic research landscape, and point to future research opportunities on the interfaces among operations, finance, and risk management. We therefore undertake an exhaustive search of related papers in the publication database. That search incorporates articles from peer-reviewed journals in the fields of operations management or corporate finance. The referred journals may be oriented toward conceptual framing, quantitative modeling, or empirical research, so in this way we cover both theoretical and practice-oriented developments in research on the operations–finance interface. For the purpose of this review, we take three steps—as described next—to identify well-cited and high-impact papers, group extant research into eight streams, relay insights from current work, and posit directions for future research.

Table 8.1 Overview of research syntheses on operations–finance interface and new contributions

Article type covered			Research scope		Conceptual synthesis	Qualitative review	Quantitative analysis	References
Conceptual	Analytical	Empirical	Integrated operations-finance risk management	Supply chain finance				
	✓		✓	✓	✓			Kouvelis et al. (2012)
	✓		✓		✓			van Mieghem (2012)
	✓		✓		✓			Bandaly et al. (2012)
	✓	✓	✓			✓		Bandaly et al. (2013)
	✓	✓	✓	✓	✓			Birge (2015)
	✓		✓	✓	✓	✓		Zhao and Huchzermeier (2015)
✓	✓	✓		✓	✓	✓		Gelsomino et al. (2016)
✓	✓	✓	✓	✓	✓	✓	✓	This chapter

Data collection. To ensure the comprehensiveness of our data set, we employ a "title, abstract, keywords" search of several frequently used keywords in the Scopus database.[1] Thus we search for the strings "operations finance interface", "supply chain finance (or financing)", "financial supply chain (management)", "capital constraint supply chain", "trade credit", "inventory financing", "(reverse) factoring", "supplier finance (or financing)", "purchase order finance (or financing)", "capacity investment budget constraint", "operational and financial hedging", "working capital management (or optimization)", "operations strategy financial performance", etc.[2]; we then merge the retrieved data by removing duplications. To concentrate on peer-reviewed journal articles, we specify the document type as "article" or "review" and thereby exclude conference papers and book chapters. These initial steps yield a data set of 487 papers. Papers whose content is unrelated (or only marginally related) to the OFI are discarded, as are publications in languages other than English. This pruning leads to a sample of 242 papers. Given that most terminology in this field has emerged in recent decades, some groundbreaking early papers[3] that had a significant impact on subsequent literature might not be categorized in Scopus under the keywords just listed, but they should certainly be included in our data set. We therefore retrieve the references from the literature reviews listed in Table 8.1 and add the (non-duplicate) entries to our data set. This data collection process—which was completed on January 1, 2017—resulted in 258 entries of publication information including title, authors, affiliations, abstract, keywords, references, and number of citations.

Bibliometric analysis. To illustrate the quantity of OFI-related publications over time, Fig. 8.1 plots the number of papers published from 1958 (earliest publication year) through 2016. That quantity has grown steadily (albeit with minor fluctuations) since the early 1990s, with a significant increase in the past decade. We identify that 100 journals have published articles in our data set of research on the operations–finance interface. The top five contributing journals (in terms of paper quantity) published 97 articles, or 37.6% of the entire data set. To investigate our data set further, we conduct citation and PageRank analyses to identify the popular and prestigious articles on the OFI. This bibliometric analysis yields statistics of influential articles on each research topic and their impact in OFI and related fields.

---

[1]Scopus (https://www.elsevier.com/solutions/scopus) is the largest abstract and citation database of peer-reviewed literature that enables filtering of publication data by document type (e.g., journal article, conference paper, review, book chapter). In contrast, Google Scholar is more comprehensive but includes all types of publications without filtering by document type. The Web-of-Science database incorporates only ISI-indexed journals and so its range is narrower than that of Scopus.

[2]The term "integrated risk management" is frequently used in OFI papers; however, it is seldom listed as a keyword owing to its multiple meanings (in the absence of a rigorous definition).

[3]Early research on capital structure—for example, Modigliani and Miller (1958) and Myers (1984)—laid the foundations for further exploration of the operations–finance interface and is therefore often cited by subsequent work in this field. Yet neither of those papers would be captured by a search based solely on our listed keywords.

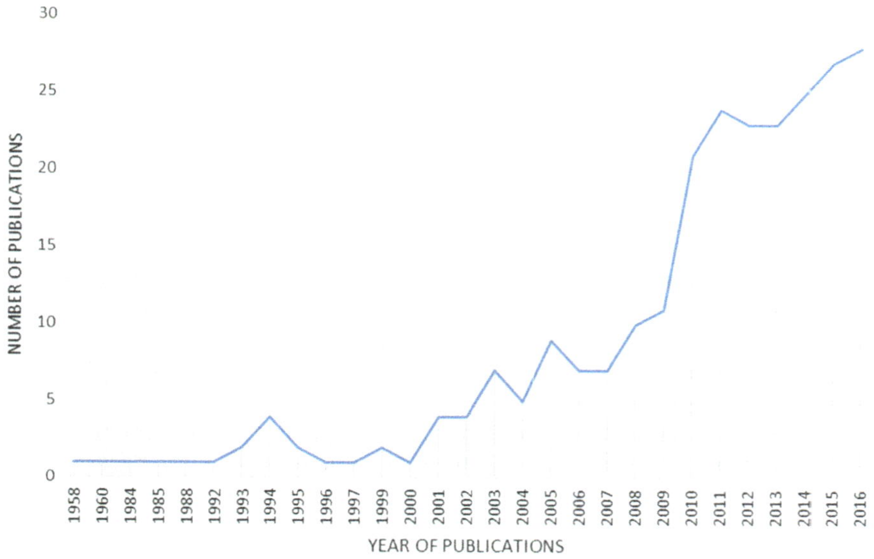

Fig. 8.1 Trend in publication of research on operations–finance interface

Research categorization. Based on a paper-citation network analysis, we classify research on the operations–finance interface into eight streams. The leading articles (i.e., those with the highest PageRank values) and the most recent papers in each stream are then summarized to provide an overview of how the literature on this topic has evolved and of current developments. Mature and emergent research streams are determined with the aid of statistical analysis, and potential directions for future exploration are proposed based on the current research landscape.

8.3 Identifying Popular and Prestigious Papers

In this section, we use citation and PageRank analyses to measure the popularity and prestige of articles on the operations–finance interface. Following the approach of Cronin and Ding (2011), we measure a paper's *popularity* by "the number of citations by other papers" and its *prestige* by "the number of citations by highly cited papers".

8.3.1 Citation Analysis

In our 258-paper data set, 154 papers (59.7%) have been cited "locally"—that is, by other papers in the data set. These local citations reflect a paper's impact within the OFI field. By "global" citations we refer to the overall number of citations in the Scopus database; these citations are indicative of a paper's impact within the broader academic community (i.e., incorporating other related fields).

Table 8.2 reports the top ten papers in terms of local citations and global citations. This table reveals that the most frequently cited papers on the operations–finance interface are the seminal articles that established the conceptual foundations and/or demonstrate methodological breakthroughs in the corresponding research streams. For instance, Babich and Sobel (2004) and Buzacott and Zhang (2004) are the first papers providing analytical efforts to breach the previously independent "silos" of operational and financial functions; they do this by incorporating initial public offerings (IPOs) and asset-based financing into production, inventory, and capacity decisions. Huchzermeier and Cohen (1996) develop the concept of operational hedging based on real options, which has since been employed (in combination with financial hedging) by Gaur and Seshadri (2005), Caldentey and Haugh (2006), and Ding et al. (2007) in studies of how to mitigate correlated operational and financial risks. Pfohl and Gomm (2009), Randall and Farris (2009), and Protopappa-Sieke and Seifert (2010) propose frameworks for supply chain finance and working capital management based on analytical and numerical

Table 8.2 Top ten papers by local and global citations

Reference	Journal	Local citation	Global citation
Buzacott and Zhang (2004)	Management Science	25	128
Huchzermeier and Cohen (1996)	Operations Research	17	197
Gaur and Seshadri (2005)	Manufacturing and Service Operations Management	17	71
Pfohl and Gomm (2009)	Logistics Research	17	19
Randall and Farris (2009)	International Journal of Physical Distribution and Logistics Management	16	23
Ding et al. (2007)	Operations Research	15	68
Caldentey and Haugh (2006)	Mathematics of Operations Research	12	32
Protopappa-Sieke and Seifert (2010)	European Journal of Operational Research	12	24
Babich and Sobel (2004)	Management Science	11	57
Klapper (2006)	Journal of Banking and Finance	11	26

approaches, and Klapper (2006) provides empirical evidence for the adoption of various supply chain finance instruments. The impact of a paper within the OFI field varies as a function of the originality and profoundness of scholars' contributions. The gap between the numbers of global and local citations indicates that papers such as Buzacott and Zhang (2004) and Huchzermeier and Cohen (1996) not only are well received in the field of operations–finance interactions but also influence related disciplines in the broader academic community.

8.3.2 *PageRank Analysis*

Whereas the citation analysis reveals the popularity of papers, a PageRank[4] analysis indicates their relative prestige. Suppose that paper A is cited by n other papers (i.e., by $p_1, \ldots, p_n$), and let $C(p_i$ denote the number of citations of paper i. Then the PageRank value (PR) of paper A is calculated as follows:

$$\mathrm{PR}(A) = (1-d)\frac{1}{N} + d\sum_{i=1}^{n}\frac{\mathrm{PR}(p_i)}{C(p_i)};$$

here d denotes a damping factor,[5] which ranges between 0 and 1 to represent the fraction of random walks that the citations continue to generate. In a network analysis of paper citations, a paper's reference entries typically follow an average length of 2 (Chen et al. 2007a); thus $1/2 = 0.5 = (1-d)$, so $d = 0.5$ is chosen for analyzing a citation network. A high PageRank value implies that a paper has been cited by many influential (i.e., frequently cited) papers. While articles with high overall citation numbers have a large impact in general, those that are often cited by *other* high-impact papers are typically of considerable importance in the focal research field (cf. Ding et al. 2009).

The top ten papers, as measured by PageRank, are listed in Table 8.3. In our data set of 258 papers, the highest PageRank value is 0.00958. Comparing Tables 8.2 and 8.3 reveals that a higher number of local and global citations may result in—but hardly guarantees—a higher PageRank value. For example: Pfohl and Gomm (2009) was cited 17 times locally and is the eighth highest by PageRank, yet Huang and Hsu (2008) achieves a slightly higher PageRank value (sixth highest) with just five local citations; the implication is that the latter work was cited by papers that were themselves relatively more influential. Four papers in Table 8.3 (Chowdhry and Howe 1999; Gaur and Seshadri 2005; Caldentey and Haugh 2006;

[4]The PageRank measure was introduced by Brin and Page (1998) to prioritize Web pages based on their "connectivity" to search engine keywords.

[5]The damping factor originally used in Google's PageRank algorithm was 0.85 (Brin and Page 1998)—based on the anecdotal observation that an Internet user typically follows about six Web pages and so the leakage probability is $1/6 \cong 0.15 = (1-d)$.

Table 8.3 Top ten papers by PageRank

Reference	Journal	PageRank	Local citation	Global citation
Gaur and Seshadri (2005)	Manufacturing and Service Operations Management	0.00958	17	71
Huchzermeier and Cohen (1996)	Operations Research	0.00836	17	197
Caldentey and Haugh (2006)	Mathematics of Operations Research	0.00727	12	32
Ding et al. (2007)	Operations Research	0.00710	15	68
Froot et al. (1993)	Journal of Finance	0.00582	8	714
Huang and Hsu (2008)	International Journal of Production Economics	0.00573	5	71
Chowdhry and Howe (1999)	European Finance Review	0.00551	7	42
Pfohl and Gomm (2009)	Logistics Research	0.00548	17	19
Gupta and Dutta (2011)	European Journal of Operational Research	0.00548	7	16
Dada and Hu (2008)	Operations Research Letters	0.00528	10	60

Ding et al. 2007) focus on integrated operational and financial hedging, while another four papers (Dada and Hu 2008; Huang and Hsu 2008; Pfohl and Gomm 2009; Gupta and Dutta 2011) analyze models used to manage financial flows in supply chains. Huchzermeier and Cohen (1996) study operational hedging by real options in global supply chain network configuration, and Froot et al. (1993) examine how financial hedging by derivative instruments can secure investment opportunities.

8.4 Evolution of Research on Operations–Finance Interface

In this section, we take two steps to classify the research streams that address the operations–finance interface. First, we analyze our data set of 258 papers using Sci^2 software to extract the paper-citation network. This procedure yields 44 paper clusters of various sizes. Second, we group the major paper clusters into seven primary research streams based on topic relevance[6]; literature syntheses and stand-alone paper clusters are placed together in an eighth group. Our

[6]This step is necessary because some of the clusters are demonstrably close in topic yet are not identified as a single cluster in the paper-citation network (since articles may not cite all related works in the same stream).

Table 8.4 Statistical overview of eight research streams on operations-finance interface

Section	Research stream	Percentage (%)	Average PageRank
8.4.1	Supply chain finance	46.9	0.002378
8.4.2	Integrated operational hedging and financial flexibility	11.6	0.004158
8.4.3	Impact of operational effectiveness on financial performance	8.9	0.002083
8.4.4	Joint financing and capacity-inventory choices	5.8	0.003138
8.4.5	Real options and hedging in operations management	4.7	0.003422
8.4.6	Integrated operational and financial planning in supply chains	3.9	0.00275
8.4.7	Valuation and risk mitigation in commodity trading	2.7	0.002455
8.4.8	Research syntheses and other topics on operations-finance interface	15.5	0.00271

categorization results in eight main groups of research (cf. Birge et al. 2007; Babich and Kouvelis 2015), Table 8.4 shows the percentage distribution and average PageRank values of each research stream. The sequence of Sects. 8.4.1–8.4.7 is organized in descending order of each stream's publication quantity.

Supply chain finance is evidently the most active field within this arena, and papers addressing trade credit account for 29.8% of this stream. Because the majority of papers on SCF were published within the last decade, their full impact is yet to be seen. Published articles on integrated operational hedging and financial flexibility have (on average) the greatest impact, in large part because most papers in that area are well established and therefore figure prominently in our data set. There is a substantial amount of research on the impact of operational effectiveness on financial performance, but this stream does not exhibit high (average) PageRank values; one reason is that such research is mostly data-driven and so there are relatively few cross-citations among papers. The innovativeness of scholarly work on joint financing and capacity inventory choices is likely responsible for its high PageRank value, on average, while research on real options and hedging in operations management achieves its high average impact by providing the very foundations for joint deployment of operational and financial hedging. There are two research streams—namely, integrated operational and financial planning in supply chains and valuation and risk mitigation in commodity trading—that are mostly driven by specific applications and so have limited influence overall in the OFI field. Research syntheses and other topics related to the operations–finance interface encompass a number of literature summaries and recent innovations that are sufficiently dispersed to preclude a significant research impact.

To ensure clarity and consistency, each of the first two research streams are divided (in Sects. 8.4.1 and 8.4.2, respectively) into three subgroups by methodology[7]; each of the other streams, which we detail in Sects. 8.4.3–8.4.7, has primarily adopted one methodology. Citation and PageRank measures each exhibit a strong positive correlation with a paper's age, so they favor established articles in early research. Articles published more recently usually take time to deploy their impact, in particular, paradigm-changing research that deviates from prevailing conventions typically has limited early impact yet exceptional long-term impact (Wang et al. 2013). We therefore summarize not only the top five PageRank papers[8] but also representative articles from recent work in each stream to provide an overview of how OFI research has evolved.

8.4.1 Supply Chain Finance

To present an overview on supply chain finance, we classify this stream by three methodologies: conceptual research, analytical modeling, and empirical assessment. Table 8.5 presents the top five conceptual papers on SCF in a descending order of their PageRank values.[9] As one of the fastest-growing areas of research on the operations–finance interface, supply chain finance establishes a conceptual foundation—based on principal–agent theory—while assuming that firms within and outside the supply chain have asymmetric information. Hence firms within the supply chain can serve as intermediaries to resolve the issue of asymmetric information between capital seekers and capital markets (Pfohl and Gomm 2009). Supply chain finance aims to enhance the allocation of working capital through cross-functional coordination of operational and financial departments and inter-organizational collaboration among supply chain partners (Hofmann 2005; Pfohl and Gomm 2009).

Closely related research argues that working capital performance, as measured by the cash conversion cycle, can be improved by shortening the inventory processing period, reducing accounts receivable, and increasing accounts payable. Thus the CCC can influence the return on capital employed (ROCE) directly by

---

[7]For ease of exposition, we categorize the research streams in Sects. 4.1 and 4.2 by methodology. However, research featuring multiple methodologies can be found in both streams. For example, Cai et al. (2014) and Gaur and Seshadri (2005) employ both analytical and empirical approaches in examining (respectively) supply chain finance and integrated risk management. For an overview of multi-methodological research in operations management, see Choi et al. (2016).

[8]Given the vast literature on operations–finance interface, we restrict our attention to top five PageRank papers in each stream to provide a navigation of the seminal research for brevity and consistency.

[9]Tables 8.6, 8.7, 8.8, 8.9, 8.10, 8.11, 8.12, 8.13, 8.14 and 8.15 are likewise presented in a descending order of the papers' PageRank values.

Table 8.5 Top five PageRank papers featuring conceptual research on supply chain finance

Concept of supply chain finance	Actors	Objects	Levers	Managerial insights	References
"Supply chain finance is the inter-company optimization of financing as well as the integration of financing processes with customers, suppliers, and service providers in order to increase the value of all participating companies."	Suppliers, customers, the focal company, financial intermediaries, and logistics service provider	Fixed assets, working capital	Volume, duration, and capital cost rate	Based on the principal-agent theory, "SCF turns the actors within the supply chain into intermediaries who can partly overcome the problem of asymmetric information between capital markets and the parties seeking capital."	Pfohl and Gomm (2009)
"Supply chain finance is an approach for two or more organizations in a supply chain, including external service providers, to jointly create value through means of planning, steering, and controlling the flow of financial resources on an inter-organizational level."	Shipper or sender, and its supplier and costumer, logistics service provider, financial institution, investors	Supply chain performance indicators on financial goals	Investment, financing, and accounting and the functions of procurement, production, and sales. The cash outlay for the investment and the value obtained from the investment	"Procurement, production, sales, and recycling can potentially free up internal financing resources." "SCF… improves the debt financing alternatives and conditions for all collaboration partners."	Hofmann (2005)
Supply chain financial management examines "how firm financial management techniques may be used to improve overall supply chain profitability and performance."	Collaborating firms across the supply chain	Inventory, accounts receivable, accounts payable, revenue and cost of goods sold (COGS)	Cash flow, payback leverage points, shifting inventory, differing cost of capital	"Reduce inventories held; reduce accounts receivable by having customers pay faster; and extend accounts payable by taking longer to pay suppliers."	Randall and Farris (2009)

(continued)

Table 8.5 (continued)

Concept of supply chain finance	Actors	Objects	Levers	Managerial insights	References
Cash conversion cycle measures "the number of days between paying for raw materials and getting paid for products, as calculated by inventory days of supply plus days of sales outstanding minus average payment period for material."	The focal firm and supply chain partners	Assets and liabilities, production cycles, inventory, accounts receivable, accounts payable	Measures of liquidity and value, inbound material activities, manufacturing operations, outbound logistics, and sales	"Extend average accounts payable, shorten production cycle to reduce inventory days of supply, reduce average accounts receivable."	Farris and Hutchison (2002)
"Supply chain finance is an integrated approach that provides visibility and control over all cash-related processes within a supply chain."	Firms within the supply chain	Current assets and current liabilities including inventory, accounts receivable and accounts payable	Cash conversion cycle, operating income, and return on capital employed (ROCE)	"Achieving transparency and analyzing their benchmarks, focusing on the leverage factors, determining the optimal CCC level."	Grosse-Ruyken et al. (2011)

employed capital and indirectly through operating income (Farris and Hutchison 2002; Randall and Farris 2009; Grosse-Ruyken et al. 2011).

Most analytical explorations on supply chain finance, which are summarized in Table 8.6, are based on the Modigliani–Miller theorem, which posits the independence of operations and finance under the strong assumption of a perfect capital market (i.e., rational investors and the *absence* of asymmetric information, incentive misalignment, corporate taxes, and transaction costs; see Modigliani and Miller 1958). When market frictions and various information structures are considered, the interconnections between operations and finance in supply chains motivate the examination of numerous supply chain finance instruments.

Trade credit,[10] a SCF topic drawing considerable research interest, includes papers that examine the joint optimization of supply chain decisions involving contract terms on payment extension and/or early discounts when the capital-constrained buyer seeks working capital financing. The effect of trade credit terms on order quantity and frequency, supply chain profitability, and the resulting interactions with bank finance is a common research theme (see e.g. Huang and Hsu 2008; Kouvelis and Zhao 2012), and recent advances focus on the interactions between trade credit and supply chain–coordination contracts as quantity discounts, buybacks, two-part tariffs, and revenue sharing (Lee and Rhee 2010; Kouvelis and Zhao 2016; Xiao et al. 2017). Additional types of trade finance instruments (e.g., vendor financing and short-term financing) have been investigated in connection with market variables that include the interest rate, salvage price, and competition structure (Brennan et al. 1988; Raghavan and Mishra 2011).

Empirical research on supply chain finance, as shown in Table 8.7, provides industry-based evidence for the adoption of trade finance instruments and working capital management. Supply chain finance programs are usually initiated by established buyers or suppliers, financial institutions (banks), or specialized service entities such as logistics service providers; these programs aim to provide financial assistance for small- and medium-sized enterprise (SME) suppliers or buyers in need of working capital (Hofmann 2005; Pfohl and Gomm 2009; Hofmann and Kotzab 2010). It follows that working capital positions in the supply chain are the primary antecedents of SCF adoption and also determine the types of SCF instruments adopted. A variety of supply chain finance instruments—such as trade credit, factoring, reverse factoring, purchase order financing, buyer credit, inventory/work-in-progress financing, letters of credit, and open account credit—can be employed to reduce liquidity risk and enhance working capital performance (Fisman and Love 2003; Klapper 2006; Wuttke et al. 2013a, b; Vliet et al. 2015).

When *pre*-shipment finance instruments are employed, the buyer's cash flow risk increases while the supplier's shortage risk decreases; adoption of *post*-shipment finance instruments mitigates the propagation of liquidity risks in the supply chain (Wuttke et al. 2013b). Market factors including competition intensity, interest

[10]For research overviews of this topic, please refer to Chang et al. (2008) and Seifert et al. (2013).

Table 8.6 Top five PageRank papers featuring analytical exploration of supply chain finance

Supplier	Buyer	Financial strategy	Market factors	Primary results	References
A supplier offers the full trade credit to the buyer	A buyer offers the partial trade credit to his/her customer	Trade credit	Annual total relevant cost, optimal cycle time of total relevant cost, optimal order quantity	"When the customer's fraction of the total amount owed payable at the time of placing an order offered by the retailer is increasing, the retailer will order less quantity and increase order frequency." "When the customer's trade credit period offered by the retailer is increasing, the retailer will order more quantity to accumulate more interest to compensate the loss of interest earned when longer trade credit period is offered to his/her customer…[under a certain condition]"	Huang and Hsu (2008)
A seller borrows from bank to provide financial intermediary to buyer	A buyer receives long-term credit from the seller	Vendor financing	Interest rate, customer wealth, duopoly competition	"…vendor financing may be optimal for a firm when demand is less elastic in the credit market than in the cash market because of adverse selection and when the reservation prices of credit customers are systematically lower than those of cash customers… in oligopolistic markets, vendor financing can be used to reduce competition since some firms can concentrate on the credit market while other firms maintain a larger market share in the cash market."	Brennan et al. (1988)
A manufacturer borrows from a lender	A retailer borrows from a lender	Short-term financing	Salvage prices, interest rates, and rate of production of defective quantity	"…if either the retailer or manufacturer has sufficiently low level of cash, a joint decision is beneficial for the lender as well as for both the borrowing firms."	Raghavan and Mishra (2011)

(continued)

Table 8.6 (continued)

Supplier	Buyer	Financial strategy	Market factors	Primary results	References
A capital-constrained agent offers trade credit to buyer	A capital-constrained agent receives trade credit	Inventory financing: trade-credit and bank finance	Coordination contracts: all-unit quantity discount, buyback, two-part tariff, and revenue sharing	"Positive financing costs call for trade-credit in order to subsidize the retailer's costs of inventory financing. Using trade-credit in addition to markdown allowance, the supplier fully coordinates the retailer's decisions for the largest joint profit, and extracts a greater portion of the maximized joint profit."	Lee and Rhee (2010)
A supplier offers trade credit to the buyer	A buyer of newsvendor type under capital constraint	Trade credit	Order quantity, trade credit interest rate, wholesale price	"…optimally structured trade credit contracts have the supplier offering interest rates [that]…give(s) incentives to the retailer to pay up front his order using his working capital to the extent possible. Thus, open account financing, with zero trade credit rate, is among the potentially optimal trade credit contracts. The result implies that optimally priced trade credit is cheaper than bank credit."	Kouvelis and Zhao (2012)

Table 8.7 Top five PageRank papers on empirical assessment of supply chain finance

Sample size	Sample region	Unit of analysis	Observation period	Methodology	Financial strategies	References
479	48 countries, 25 high-income and 23 middle-income countries	Country	1993–2003	Regression, case study	Factoring, reverse factoring, purchase order financing	Klapper (2006)
Approxim. 7300 firms	US, Europe	Firms	1995–2004	Case study, conceptual model	Working capital management by cash-to-cash cycle	Hofmann and Kotzab (2010)
36 observations	43 countries	Countries and firms	1980–1990	Regression	Trade credit	Fisman and Love (2003)
8 cases based on 40 interviews	Europe	Firms	2012	Case study	Buyer credit, inventory/ work-in-progress financing, reverse factoring, letters of credit, open account credit, bank loan	Wuttke et al. (2013b)
6 cases	Europe	Firms	2012	Case study	Adoption process of supply chain finance	Wuttke et al. (2013a)

rates, receivables quantity, and the supplier's working capital objectives determine the benefits (and thereby the adoption, or not) of reverse factoring for supply chain partners (Wuttke et al. 2013a; Iacono et al. 2015).

8.4.2 *Integrated Operational Hedging and Financial Flexibility*

Analogously to our approach in Sect. 8.4.1, here we categorize research on integrated risk management of operational hedging and financial flexibility into three subgroups by methodology. Conceptual research on integrated risk management has proposed frameworks for joint optimization of operational strategies and financial instruments; see Table 8.8. Integrated risk management features the joint analysis of operational and financial exposure to uncertainty as well as on synchronized risk mitigation through operational hedging and financial flexibility (Miller 1992; Trigeorgis 1993; Triantis 2000; Meulbroek 2002; Kouvelis et al. 2012). On the one hand, operational hedging involves decisions concerning risk preferences and exposure of the focal firm and its supply chain partners (Meulbroek 2002); the real options to defer, expand, contract, abandon, or alter strategies when more information can be obtained (Trigeorgis 1993; Triantis 2000); and risk mitigation strategies that include avoidance, control, cooperation, limitation, and flexibility (Miller 1992); or buffering, pooling, and contingency planning (Kouvelis et al. 2012). On the other hand, financial risk management includes the hedging of

Table 8.8 Top five PageRank papers featuring conceptual research on integrated risk management

Concept of integrated risk management	Operational hedging strategies	Financial flexibility instruments	Application context	Reference
"Integration of risks and integration of risk management ways" "to choose the optimal level of risk to maximize shareholder value"	Operational risk management "is not only a decision about how much risk the firm should bear, it is also a decision about how much risk the firm's customers or suppliers are prepared to bear"	Hedging aggregate risk of loss by derivatives or insurance, risk adjustment via the capital structure	Multinational corporations (MNCs)	Meulbroek (2002)
Joint financial and real options "alternative mechanisms of limiting downside risk while allowing the upside to be exploited"	Real options refer to "the ability to delay making decisions until an opportunity to obtain more information"	The use of financial derivatives	MNC, project management	Triantis (2000)
Interaction between real options and financial flexibility	Management can use real options "to defer, expand, contract, abandon, or otherwise alter a project at different stages during its useful operating life"	Financial flexibility options including equity holders' option to default on debt payments deriving from limited liability, lenders' option to abandon via staged debt financing, mixed debt-equity venture capital financing	Project investment and management	Trigeorgis (1993)
Integrated risk management with strategic and financial responses to uncertainties	Strategic risk management by avoidance, control, cooperation, limitation, and flexibility	Financial risk management by derivatives and insurance	International management	Miller (1992)
Integrated operations-finance risk management with risk identification, assessment, and mitigation	Operational risk mitigation strategies by buffering, pooling, and contingency planning	Financial hedging strategies and supply chain finance	Action-based supply chain risk management	Kouvelis et al. (2012)

aggregate risk exposure by financial derivatives and insurance contracts, risk adjustment by way of altering the firm's capital structure (Miller 1992; Triantis 2000; Meulbroek 2002; Kouvelis et al. 2012), and such financial flexibility options as equity holders' option to default on debt payments, lenders' option to abandon a commitment via staged debt financing, and venture capital financing by a mix of debt and equity (Trigeorgis 1993).

The analytical approach to integrated risk management has mainly focused on examining the relationship between operational flexibility and financial hedging and in mitigating their interrelated uncertainties; papers in this group are summarized in Table 8.9. The mainstream of the analytical literature (represented by Mello et al. 1995; Chowdhry and Howe 1999; Hommel 2003; Ding et al. 2007; Chen et al. 2014) concentrates on integrated operational flexibility and financial hedging under exchange rate and demand risks; the mitigation of demand risk by joint financial hedging and inventory management is analyzed based on the case where empirical evidence shows that demand is correlated with a financial asset price (Gaur and Seshadri 2005). The joint optimization of financial hedging and operational decisions of a corporation are explored when profits are correlated with financial market returns (Caldentey and Haugh 2006)—and of a two-firm supply chain when the retailer is budget constrained (Caldentey and Haugh 2009). Product flexibility and financial hedging are shown to be complements (resp. substitutes) when the demand for different products is positively (resp. negatively) correlated; postponement flexibility and financial hedging are substitutes (Chod et al. 2010). In the presence of financial market frictions, liquidity makes a major contribution to risk management while operational and financial hedging exhibit substitution effects (Gamba and Triantis 2014).

Most of the research in this stream sheds light on the complementarity between operations and finance, although substitution effects are also evident. Operational hedging mainly drives profitability, and financial hedging (when adopted jointly) tailors cash flow variability. Thus the primary emphasis has been on the coordination and collaboration between operations and finance in a centralized approach to integrated risk management. From a technical viewpoint, variance-based risk measures—including mean-variance models (Chowdhry and Howe 1999; Hommel 2003; Ding et al. 2007; Chen et al. 2014)—treat upside potential and downside risk symmetrically. As an alternative, more recent research efforts are devoted to such downside risk measures as value at risk (Park et al. 2017) and conditional value at risk (Koberstein et al. 2013, Chap. 5) Zhao and Huchzermeier (2017).

Empirical research on integrated risk management has tested the relationship between—and the relative effectiveness of—operational and financial hedging based on industrial evidence (see Table 8.10). In this context, operational hedging features geographic dispersion of multinational corporations' global subsidiary network (Allayannis et al. 2001; Kim et al. 2006), fuel pass-through agreements and charter operations in airline industry (Carter et al. 2006), expected changes in operational volatility due to acquisitions (Hankins 2011), and real options; the latter include market entry and exit, production and sourcing switches, and the acquisition and sale of a subsidiary (Aabo and Simkins 2005). Financial hedging has been

Table 8.9 Top five PageRank papers featuring analytical exploration of integrated risk management

Source of uncertainty			Interdependence			Operational strategies	Financial instruments	Relationship analysis		Optimization approach		References
Demand	FX	Others	Alternative	Constraint	Correlation			Complements	Substitutes	Centralization	Decentralization	
✓		✓			✓	Inventory mitigation	Financial assets, options	✓		✓		Gaur and Seshadri (2005)
✓		✓			✓	Newsvendor inventory	Financial assets	✓		✓		Caldentey and Haugh (2006)
✓	✓		✓		✓	Capacity plan, allocation options	Currency forwards, options	✓		✓		Ding et al. (2007)
✓	✓		✓		✓	Production flexibility	Currency forwards, options	✓		✓		Chowdhry and Howe (1999)
✓	✓		✓	✓	✓	Production flexibility	Currency bonds, swaps, forwards	✓			✓	Mello et al. (1995)

Table 8.10 Top five PageRank papers on empirical assessment of integrated risk management

Sample size	Sample region	Observation period	Operational strategies	Financial instruments	Complements	Substitutes	**References**
424	US	1996–2000	Geographic dispersion	"Total notional amount of currency derivatives divided by total foreign activities"	✓	✓	Kim et al. (2006)
756	US	1996–1998	Geographic dispersion	Foreign-currency derivatives or foreign debt	✓		Allayannis et al. (2001)
259	US	1992–2003	"Fuel pass-through agreements, and the use of charter operations"	Financial hedging by derivatives on fuel prices	✓		Carter et al. (2006)
487	US	1995–2003	"Expected percentage change in operational volatility due to the acquisition"	Use of derivatives for interest rate hedging		✓	Hankins (2011)
52	Denmark	2001	Real option actions including market entry/exit, production/sourcing changes, acquisition/sales of foreign company	Financial hedging of currency risk		✓	Aabo and Simkins (2005)

measured by the magnitude of foreign currency derivatives or foreign debt, where this hedging is used to manage exposure to such factors as the interest rate, price, and currency risks (Allayannis et al. 2001; Aabo and Simkins 2005; Carter et al. 2006; Kim et al. 2006; Hankins 2011).

The complementary relationship between operational and financial hedging has been supported by the finding that operational hedging does not effectively substitute financial hedging (Allayannis et al. 2001). Hence firms that engage in relatively more operational hedging are likely to adopt financial derivatives, and such joint operational and financial hedging is associated with reduced risk exposure and enhanced firm value (Carter et al. 2006; Kim et al. 2006). Operational hedging is typically used to mitigate long-term economic exposure, whereas financial hedging is more often employed to manage short-term transaction exposure (Kim et al. 2006). The claim that operational and financial risk management can be substitutes is supported by evidence that the use of real options (resp., financial hedging) increases (resp., decreases) with the length of the time horizon (Aabo and Simkins 2005). Moreover, operational hedging by acquisitions (that reduce operational volatility) is a viable substitute for financial hedging by derivatives (Hankins 2011).

8.4.3 Impact of Operational Effectiveness on Financial Performance

Supply chain operations constitute the backbone of financial performance. The papers listed in Table 8.11 analyze empirically the intricate connections between operational practices—such as supply chain integration, membership, and glitches as well as just-in-time (JIT) inventory—and customer service, shareholder value, and profitability measures; these measures include return on assets (ROA), return on investment (ROI), return on sales (ROS), and profit margin. The relationship between supply chain integration and financial metrics is fully mediated by customer service (Vickery et al. 2003). Supplier integration has a positive effect on firm performance, while system integration has a greater effect on financial metrics than does process integration (Huo et al. 2013). The main effects of concentrated supply chain membership are higher profits for downstream partners, and increased productivity (more so than increased efficiency) is the main driver of ROA (Lanier et al. 2010). Supply chain glitches are related to declines in shareholder wealth as measured by abnormal stock returns and also to reduced firm performance as measured by operating income, ROA, and ROS (Hendricks and Singhal 2003, 2005). Maximizing the productivity of inventory is positively related to subsequent stock performance because portfolio returns depend on the timely consideration of inventory information when making stock investment decisions (Alan et al. 2014).

Table 8.11 Top five PageRank papers assessing impact of operational effectiveness on financial performance

Sample size	Sample region	Observation period	Methodology	Operational perspectives	Financial performance	Primary results	References
57	North America	2001	Structural equations modeling	Supply chain integration	Customer service, pre-tax return on assets, return on investment, return on sales	"The relationship of supply chain integration to financial performance was indirect, through customer service; i.e., customer service was found to fully (as opposed to partially) mediate the relationship between supply chain integration and firm performance for first tier suppliers in the automotive industry."	Vickery et al. (2003)
519	Global	1989–2000	Regression	Supply chain glitches	Shareholder value as abnormal stock returns	"Supply chain glitch announcements are associated with an abnormal decrease in shareholder value of 10.28%…larger firms experience a less negative market reaction, and firms with higher growth prospects experience a more negative reaction."	Hendricks and Singhal (2003)
276	Global	1980–1996 and 1997–2006	Regression	Concentrated supply chain membership	Return on assets, margin, turnover and cycle	"…the profitability benefits of supply chain relationships are captured predominantly by downstream chain members, whereas cash cycle benefits are realized throughout the supply chain…chain members' financial performance varies systematically with measures of downstream bargaining power, downstream relationship duration, and degree of supply consolidation."	Lanier et al. (2010)
617	China	2011	Exploratory factor analysis	Supplier integration	Growth in sales, return on sales, profit, market share, return on investment	"…normative and mimetic pressures are positively related to both system and process integration; while coercive pressures are only positively related to process integration and	Huo et al. (2013)

(continued)

Table 8.11 (continued)

Sample size	Sample region	Observation period	Methodology	Operational perspectives	Financial performance	Primary results	References
						not significantly related to system integration…both system and process integration have a positive impact on financial performance.”	
253	US	1996–1997	Regression	Just-in-time (JIT)	Return on assets, return on sales, and cash flow margin	“From both a cross-sectional and longitudinal perspective, …[there are] significant statistical relationships between measures of profitability and the degree of specific JIT practices used…firms that implement and maintain JIT manufacturing systems will reap sustainable rewards as measured by improved financial performance.”	Fullerton et al. (2003)

8.4.4 Joint Financing and Capacity Inventory Decisions

In contrast to supply chain finance that examines *multiple* firms' decisions to coordinate channel-wide material and monetary flows, most models of joint financing and the capacity inventory decision analyze the interactions between capital structure and operations strategy in a *single*-firm setting. The majority of this research, which we summarize in Table 8.12, analyzes and optimizes the interconnections between operational strategies in capacity, production, and inventory planning, on the one hand, and, on the other hand, financial instruments such as loan, asset-based financing, initial public offerings, and supplier subsidies (Babich and Sobel 2004; Buzacott and Zhang 2004; Dada and Hu 2008; Babich 2010; Chod and Zhou 2014). The Modigliani–Miller theorem's validity is frequently tested by proposing conditions under which financing and operational decisions are independent (e.g., Babich 2010) and by identifying the factors that render these two functions inseparable (e.g., Babich and Sobel 2004; Buzacott and Zhang 2004). The proposed jointly optimal decisions are intended to adjust loan terms and capital structure toward the end of better matching supply with demand in both single- and multi-product settings (Dada and Hu 2008; Chod and Zhou 2014).

The financing instruments employed typically aim to release the budget constraint imposed on operational investment and thereby modify the interactions between decisions that affect capacity inventory and financing. For instance, order quantities and supplier subsidies are substitutes for the manufacturer in a proportional random-yield model, and their relationship is determined by the convexity of the manufacturer's cost function in a model with up-front costs (Babich 2010). Furthermore, an externally financed firm should invest more in flexible capacity than should an equity-financed firm, since resource flexibility ensures risk-pooling benefits and reduces financing costs via more favorable credit terms (Chod and Zhou 2014). In this setting, a firm operating on a secured loan should be more in favor of increasing flexible capacity than a firm with an unsecured loan—irrespective of credit market competitiveness (Boyabatli and Toktay 2011). A firm with less financial flexibility should adopt dedicated capacities, across a wide range of unit investment costs, unless the capacity budget is severely constrained (Boyabatli et al. 2016).

8.4.5 Real Options and Hedging in Operations Management

The use of real options for operational hedging is based on the application of financial options logic to supply chain and operations management; papers in this vein are described in Table 8.13. Various types of real options–based operational flexibility to postpone, expand, contract, exit, switch or improve operational decisions can be adopted to mitigate sources of uncertainty, which include exchange rate volatility, demand risk, and unreliable supply (Kogut and Kulatilaka 1994; Huchzermeier and Cohen 1996; Kazaz et al. 2005; van Mieghem 2007; Wang et al. 2010).

Table 8.12 Top five PageRank papers on joint financing and capacity inventory decisions

Focal firm	Operational strategies	Financial instruments	Major factors	Primary results	References
A manufacturer sourcing from risky suppliers	Capacity reservation	Supplier subsidy	Volatility of supplier asset level, demand distribution	The manufacturer (buyer) can make ordering decisions independent of subsidy decisions under the assumptions of independent supply and demand shocks, no inventory, random capacity, zero upfront costs. The optimal subsidy policy has a "subsidize-up-to" structure and the optimal ordering decisions show newsvendor fractiles under certain conditions.	Babich (2010)
A capital-constrained newsvendor	Order quantity	Bank loan	Interest rate, coordinating loan schedules	"The lender charges an interest rate that decreases in the newsvendor's equity…a non-linear loan schedule that coordinates the channel [is derived]".	Dada and Hu (2008)
A manufacturer of a single product	Production and inventory control	Asset-based financing	Interest rate, loan limits, cash position, bankruptcy risk, demand uncertainty	"While an asset-based loan allows a firm to continue growing without having to renegotiate the loan, the equivalent unsecured loan results in more earnings early on but limits earnings later." "…retailers are able to enhance their cash return over what it would be if the retailers only used their own capital."	Buzacott and Zhang (2004)
A start-up prior to initial public offering (IPO)	Capacity expansion	Initial public offering	Production capacity, demand distribution, risk-free discount factor, current assets, bankruptcy risk	"…operational and financial decisions are inseparable", "…the amount of the bank loan [are derived] as a function of capacity increment and state variables…the firm should borrow exactly the shortfall	Babich and Sobel (2004)

(continued)

Table 8.12 (continued)

Focal firm	Operational strategies	Financial instruments	Major factors	Primary results	References
				between available current assets and the financial resources required for production and capacity expansion."	
A firm investing in flexible and nonflexible capacities for two products	Resource flexibility	Debt financing	Debt and interest value, agency cost of debt, tax benefit of debt, cost of issuing equity, cost of financial distress, demand states	"…in the presence of debt, resource flexibility has benefits in addition to reducing the mismatch between supply and demand. Namely, resource flexibility mitigates the shareholder–debtholder agency conflict as well as the risk of costly default…resource flexibility mitigates the underinvestment problem because it reduces the probability that a firm will go bankrupt with some of its capacity being fully utilized."	Chod and Zhou (2014)

Table 8.13 Top five PageRank papers on real options and hedging in operations management

Source of uncertainty	Focal firm	Operational (and financial) strategies	Major factors	Primary results	References
Currency risk	A global firm with global production network	Operational hedging by supply chain network options and financial hedging by forward contracts	Exchange rates and their correlations, demand states, switching costs, market prices	"…the firm's global manufacturing strategy determines options for alternative product designs as well as supply chain network designs. Product options introduce international supply flexibility. Supply chain network options determine the firm's manufacturing flexibility through production capacity and supply chain network linkages."	Huchzermeier and Cohen (1996)
Demand risk, profit variability risk	A risk-averse newsvendor that produces two products	Diversification by dedicated resources, resource sharing and flexibility, demand pooling	Risk exposure, risk aversion, demand correlation	"Risk-averse newsvendors may increase network capacity and total spending above risk-neutral levels because rebalancing capacity may decrease profit variance." "Inexpensive resources supplying the lower-profit variance market may be increased for operational hedging, especially with strong negative correlations."	van Mieghem (2007)
Currency risk	A multinational corporation with global production facilities	Shifting production between two manufacturing plants located in different countries	Exchange rates, switching costs, growth options, transfer pricing	There are "hysteresis effects and within-country growth options…the management of cross-border coordination has led to changes in the heuristic rules used for performance evaluation and transfer pricing".	Kogut and Kulatilaka (1994)

(continued)

Table 8.13 (continued)

Source of uncertainty	Focal firm	Operational (and financial) strategies	Major factors	Primary results	References
Supply risk, demand uncertainty	A buyer can source from multiple suppliers	Dual sourcing and process improvement to increase supplier reliability	Demand state, procurement quantity, supplier reliability, cost difference between suppliers, unit improvement cost	"For random capacity, improvement is increasingly favored over dual sourcing as the supplier cost heterogeneity increases, but dual sourcing is favored over improvement if the supplier reliability heterogeneity is high. In the random yield model, increasing cost heterogeneity can reduce the attractiveness of improvement, and improvement can be favored over dual sourcing if the reliability heterogeneity is high."	Wang et al. (2010)
Currency risk, demand uncertainty	A global manufacturer makes production and allocation decisions	Production hedging, allocation hedging	Exchange rates and their correlations, timing of demand realization, price setting	"…the prevalence of production hedging is moderated by the degree of correlation between exchange rates…production and allocation hedging are robust for these generalizations and should be integrated into the overall aggregate planning strategy of a global manufacturing firm."	Kazaz et al. (2005)

As described in Sect. 8.4.1, the real options valuation and hedging framework sets the stage for integrated operational hedging and financial flexibility. The management of currency risk by way of exercising production, switching, and allocation options (Kogut and Kulatilaka 1994; Huchzermeier and Cohen 1996; Kazaz et al. 2005) is in the mainstream of operational hedging via real options in global supply chain networks. Here scholars have compared related management strategies and explored how combining them can enhance firm profitability. For instance, operational diversification, flexibility, and sharing strategies are evaluated in a newsvendor network (van Mieghem 2007), where operational hedging favors capacity with lower costs so that the market can be supplied with lower risk of profit variability. The relative effectiveness—at improving supplier reliability—of making process improvements versus flexible sourcing from multiple suppliers is examined in the context of mitigating supply risk, where the combination of dual sourcing and process improvements yields significant value when supplier reliability is low and/or demand exceeds production capacity (Wang et al. 2010). In the presence of supply chain network options, financial hedging via forward contracts on exchange rates have limited impact on operational hedging (Huchzermeier and Cohen 1996).

8.4.6 *Integrated Operational and Financial Planning in Supply Chains*

The joint planning of operational aspects and financial metrics in supply chain networks is typically analyzed in optimization, computational, and simulation models of relatively high complexity; see Table 8.14 for a list of leading papers in this stream. Hence these research efforts have been invested in proposing effective heuristics for supply chain planning (Guillen et al. 2007; Gupta and Dutta 2011) and presenting the value of a holistic approach to simultaneous optimization of operational and financial performance measures (Protopappa-Sieke and Seifert 2010; Longinidis and Georgiadis 2011, 2014).

In this context, discussion of *financial* aspects focuses on the management of working capital and liquidity (Guillen et al. 2007; Protopappa-Sieke and Seifert 2010; Gupta and Dutta 2011; Longinidis and Georgiadis 2011) and on financing schemes (Longinidis and Georgiadis 2014). At the same time, the *operational* decisions address the scheduling of capacity, production, inventory, and distribution. The value of integrated operational planning and financial budgeting is emphasized in comparison with traditional sequential approaches, under which operational decisions are made first and financial adjustments are made afterwards (Guillen et al. 2007). Under an integrated approach, managers also consider shareholder value when designing a supply chain network (Longinidis and Georgiadis 2011). Also, firms should evaluate the configuration of supply chain infrastructure with an eye toward improving their financial ratios and while considering the outlook for real estate markets (Longinidis and Georgiadis 2014).

Table 8.14 Top five PageRank papers on integrated operational and financial planning in supply chains

Source of uncertainty	Research scope	Operational aspects	Financial perspectives	Major factors	Primary results	References
Arrival of invoices and receipt of cash	A wholesaler purchases products from manufacturers and distributes to retailers	Inventory decisions	Timing of payments with early payment discount and late payment interest	Discount and interest rates, complexity and accuracy of solution approaches	The decision of "which invoice should be paid at any given point in time…is influenced by the pending invoices and the available cash…for very large problems of several years and hundreds of invoices the dynamic heuristic may be preferred over the interval heuristic."	Gupta and Dutta (2011)
Demand uncertainty	A supply chain of external and internal suppliers, batch plants, warehouses, and customers	Capacity production planning and inventory scheduling	Cash management under budgetary constraints	Inventory levels, demand forecast	"…the results obtained by the integrated model are compared with those computed by the traditional sequential strategy, in which the operations are firstly computed and the finances are fitted afterwards. The obtained results show the importance of devising broader modeling systems for SCM leading to increased overall earnings…"	Guillen et al. (2007)
Capital restrictions, payment delays	A firm determines the order quantity under working capital restrictions and payment delays	Service level, profit margin, inventory management	Payment terms and working capital requirements, return on investment	Payment delays, profit margin, service level	"Increases/decreases in the upstream/downstream payment delays favor the system's operations by decreasing operational costs. Moreover increases in the working capital	Protopappa-Sieke and Seifert (2010)

(continued)

Table 8.14 (continued)

Source of uncertainty	Research scope	Operational aspects	Financial perspectives	Major factors	Primary results	References
					employed in the system decrease the total operational cost, increase the total financial cost and lower the return on working capital investment."	
Demand uncertainty	A multi-product, four-echelon supply chain network	Planning of warehouses/ distribution centers, production, inventory, and transportation decisions	Liquidity ratios, assets management ratios, solvency ratios, profitability ratios, economic value added	Weighted average cost of capital, production, handling, transport, infrastructure, storage costs	"The modeling of financial statements enables SC [supply chain] managers to take holistic decisions without underestimating the basic objective of a profit company, which is the creation of shareholder value."	Longinidis and Georgiadis (2011)
Demand uncertainty	A multi-product, four-echelon supply chain network	Planning of warehouses/ distribution centers, production, inventory, and transportation decisions	Number and timing of warehouses and distribution centers to be sales and leaseback	Transported quantity, handling capacities, inventory, total resource availability, production capacity	"…managers should be able to evaluate how the supply chain network decisions contribute to the overall performance of the company as opposed to only cost oriented indicators. By employing advance financial management methods, such as sale and leaseback, fixed assets could be the medium to improve liquidity and strengthen credit solvency."	Longinidis and Georgiadis (2014)

8.4.7 Valuation and Risk Mitigation in Commodity Trading

Commodity risk management typically integrates the physical storage and trading of commodities (e.g., grains, fruits, vegetables, precious metals, coal, oil, and natural gas) and hedging via financial derivatives, see Table 8.15. This stream is related to the analytical research on integrated operational and financial hedging of currency risk, but it differs in that the volatility of commodity prices affects firms' procurement costs and selling prices while exchange rate volatility has an effect on foreign operations.

On the operational side, both storage capacity and inventory decisions are crucial in matching a commodity's supply with its demand. On the financial side, firms can exploit the volatility of commodity prices by using the real options of storage assets in commodity trading (Secomandi 2010a) and/or can mitigate that volatility by using financial derivatives tied to commodity prices (Kouvelis et al. 2013; Turcic et al. 2015). The valuation of commodity conversion assets, storage, and transport capacity can be computed as spread options that are contingent on commodity prices (Secomandi 2010b; Secomandi and Wang 2012). When a commodity storage asset has space and capacity limits, the operational and trading decisions are intertwined because the optimal strategy depends on both the commodity spot price and the merchant's inventory availability (Secomandi 2010a). Financial hedging can reduce the inventory of risk-averse traders in a multi-period setting. It follows that financial managers should incorporate inventory-level considerations into their financial hedging decisions, and operational hedging by procuring inputs from the spot market reduces risk more effectively than does financial hedging (Kouvelis et al. 2013). Moreover, the risk of supply disruptions could, under certain circumstance, lead to financial hedging of input costs by both members of a two-firm supply chain (Turcic et al. 2015).

8.4.8 Research Syntheses and Other OFI Topics

This category encompasses literature syntheses on the interactions between operations and finance and the clusters of papers on emerging topics that are neither subsumed by any of the primary research streams (presented in Sects. 8.4.1–8.4.7) nor sufficiently developed to constitute a new stream or field of research. Recall that representative literature syntheses were summarized in Table 8.1. Promising and innovative topics on the operations–finance interface are referenced when we discuss, in Sect. 8.5, potential directions for future research.

Table 8.15 Top five PageRank papers on valuation and risk mitigation in commodity trading

Source of uncertainty	Research scope	Operational aspects	Financial perspectives	Major factors	Primary results	References
Commodity spot price	Warehouse problem with both space and injection/ withdrawal capacity limits	Inventory management	Commodity trading	Injection/withdrawal capacity, scale factors	"…the interplay between the capacity and space limits of the storage asset and brings to light the nontrivial nature of the interface between trading and operations… mismanaging this interface can yield significant value losses. Moreover, adapting the merchant's optimal trading policy to the spot-price stochastic evolution has substantial value."	Secomandi (2010a)
Natural gas price	Pricing of natural gas pipeline capacity	Matching the supply and demand of natural gas	Capacity pricing	Volume, commodity rate, demand rate	"…capacity should be priced at its trading value, which can be interpreted as the value of a spread option on natural gas prices at two ends of a pipeline, by every player involved in natural gas transport… uncertainty in future natural gas prices is an important driver of operational performance in the pricing of this capacity."	Secomandi (2010b)
Commodity price, consumption volume	A risk-averse firm procuring a storable commodity from a spot market and a long-term supplier	Inventory management	Financial hedging	Service level, mean, variance, utility of the cash flows, correlation between demand and spot price	"…as long as futures are used in each period, alone or not, the optimal inventory policy is myopic. The optimal hedging policy, however, is never myopic, but depends on all the future optimal decisions." "Moreover…hedging may lead to inventory reduction in multi-period problems."	Kouvelis et al. (2013)
Commodity price	Network contracts for transport capacity of natural gas pipelines, energy conversion assets	Valuing capacity of commodity conversion assets, physical trading of commodities	Financially hedging the physical trading cash flows	Commodity and fuel rates, performance of the spread option heuristic	"…optimal operating policy of contracts with this flexibility is of the greedy type, whereas the one of contracts without it in general is not…greedy optimization yields a near-optimal operating policy for the latter	Secomandi and Wang (2012)

(continued)

Table 8.15 (continued)

Source of uncertainty	Research scope	Operational aspects	Financial perspectives	Major factors	Primary results	References
					type of contracts with a substantially reduced computational effort…model can also be employed to efficiently estimate the contract value sensitivities, which merchants use for financial hedging purposes."	
Stochastic input costs and demand	A risk-neutral supply chain of "selling-to-the-newsvendor" model when both firms face stochastic input costs	Order quantity and pricing decisions	Financial hedging of commodity input costs	Demand rates, order quantity, hedging policy	"…conditions under which supply chain breakdown risk will cause the supply chain members to hedge their input costs: the downstream buyer's market power exceeds a critical threshold; or the upstream firm operates on a large margin, there is a high baseline demand for downstream firm's final product, and the downstream firm's market power is below a critical threshold."	Turcic et al. (2015)

8.5 Conclusions and Future Research

This chapter presents a bibliometric analysis and also a qualitative summary in reviewing conceptual, analytical, and empirical research on the operations–finance interface. Our citation and PageRank analyses identify the lead articles by citations and PageRank in this field. We characterize the OFI research landscape by identifying eight research streams in the paper-citation network, and we summarize each stream with reference to leading papers (in terms of a PageRank metric) and representative articles from recent publications.

Opportunities for future research on operations–finance interface could be captured using one or more of the four approaches described next (Simchi-Levi 2014; Tang 2017). The first of these is *problem*-driven research, or the identification and optimization of real-world problems led by research questions that are relevant to practice. For instance, a donor (e.g., of international funds) to the private-sector supply chain may need to decide on whether to subsidize the sale or purchase of malaria drugs so they will be more affordable to customers in developing countries; in such cases, it may be suggested that the donor subsidize only purchases (and not sales). Factors affecting the donor's decision include product shelf life, pricing flexibility, retailer heterogeneity, and the supply chain structure (Taylor and Xiao 2014). The second approach is *phenomenon*-driven research—that is, the observation and summary of related innovations that could motivate new research ideas and questions. An example of this approach would be a proposal, inspired by the phenomenon of product recalls, of three solutions to address the problem of product adulteration: deferred payment for shipped product, closer inspection of the product, or both. It has been demonstrated that deferred payment can successfully prevent product adulteration by suppliers whereas inspection cannot, and combining these mechanisms does not add value (Babich and Tang 2012).

Third, *literature*-driven research consists of advancing theory and further developing previous insights in a program inspired and justified by the extant literature. For instance, research advances in the areas of product or postponement flexibility and financial hedging have led to the exploration of their relationship in a risk-averse, multi-product setting (Chod et al. 2010). Finally, *data*-driven research, or the collection and assessment of industry data, can deepen our understanding of managerial insights and test for the existence of outcomes hypothesized by theoretical models. Thus, for example, a data set of more than 2000 loans to suppliers has been used to compare pre- and post-bankruptcy financing costs (Houston et al. 2016); the authors report increased costs averaging 20% after bankruptcy that underscore the effect of a buying firm's bankruptcy on the financing costs of its major suppliers.

References

Aabo, T., & Simkins, B. J. (2005). Interaction between real options and financial hedging: Fact or fiction in managerial decision-making. *Review Financial Economics, 14,* 353–369.
Alan, Y., Gao, G. P., & Gaur, V. (2014). Does inventory productivity predict future stock returns? A retailing industry perspective. *Management Science, 60,* 2416–2434.
Allayannis, G., Ihrig, J., & Weston, J. P. (2001). Exchange-rate hedging: Financial vs. operational strategies. *American Economic Review, 91,* 391–395.
Babich, V. (2010). Independence of capacity ordering and financial subsidies to risky suppliers. *Manufacturing & Service Operations Management, 12,* 583–607.
Babich, V., & Kouvelis, P. (2015). Call for papers—Special issue of manufacturing & service operations management: Interface of finance, operations, and risk management. *Manufacturing & Service Operations Management, 17,* 271–271.
Babich, V., & Sobel, M. J. (2004). Pre-IPO operational and financial decisions. *Management Science, 50,* 935–948.
Babich, V., & Tang, C. S. (2012). Managing opportunistic supplier product adulteration: Deferred payments, inspection, and combined mechanisms. *Manufacturing & Service Operations Management, 14,* 301–314.
Bandaly, D., Satir, A., Kahyaoglu, Y., & Shanker, L. (2012). Supply chain risk management. I: Conceptualization, framework and planning process. *Risk Management, 14,* 249–271.
Bandaly, D., Satir, A., Kahyaoglu, Y., & Shanker, L. (2013). Supply chain risk management. II: A review of operational, financial and integrated approaches. *Risk Management, 15,* 1–31.
Birge, J. R. (2015). OM Forum—Operations and finance interactions. *Manufacturing & Service Operations Management, 17,* 4–15.
Birge, J., Kouvelis, P., Seppi, D. (2007). Call for papers: Special Issue of Management Science Interfaces of Operations and Finance.
Boyabatli, O., Leng, T., & Toktay, L. B. (2016). The impact of budget constraints on flexible versus dedicated technology choice. *Management Science, 62,* 225–244.
Boyabatli, O., & Toktay, L. B. (2011). Stochastic capacity investment and flexible vs. dedicated technology choice capacity investment in imperfect capital markets. *Management Science, 57,* 2163–2179.
Brennan, M., Maksimovic, V., & Zechner, J. (1988). Vendor financing. *The Journal of Finance, 43,* 1127–1141.
Brin, S., & Page, L. (1998). The anatomy of a large-scale hypertextual Web search engine. *Computer Networks ISDN Systems, 30,* 107–117.
Buzacott, J., & Zhang, R. (2004). Inventory management with asset-based financing. *Management Science, 50,* 1274–1292.
Cai, G., Chen, X., & Xiao, Z. (2014). The roles of bank and trade credits: Theoretical analysis and empirical evidence. *Production and Operations Management, 23,* 583–593.
Caldentey, R., & Haugh, M. (2006). Optimal control and hedging of operations in the presence of financial markets. *Mathematical Operations Research, 31,* 285–304.
Caldentey, R., & Haugh, M. (2009). Supply contracts with financial hedging. *Operations Research, 57,* 47–65.
Carter, D. A., Pantzalis, C., & Simkins, B. J. (2006). Does hedging affect firm value? Evidence from the US airline industry. *Financial Management, 35,* 53–86.
Chang, C. T., Teng, J. T., & Goyal, S. K. (2008). Inventory lot-size models under trade credits: A review. *Asia-Pacific Journal of Operational Research, 25,* 89–112.
Chen, L., Li, S., & Wang, L. (2014). Capacity planning with financial and operational hedging in low-cost countries. *Production of Operations Management, 23,* 1495–1510.
Chen, P., Xie, H., Maslov, S., & Redner, S. (2007a). Finding scientific gems with Google's PageRank algorithm. *Journal of Informetrics, 1,* 8–15.
Chod, J., Rudi, N., & van Mieghem, J. A. (2010). Operational flexibility and financial hedging: Complements or substitutes? *Management Science, 56,* 1030–1045.

Chod, J., & Zhou, J. (2014). Resource flexibility and capital structure. *Management Science, 60,* 708–729.

Choi, T.-M., Cheng, T. C. E., & Zhao, X. (2016). Multi-methodological research in operations management. *Production and Operations Management, 25,* 379–389.

Chowdhry, B., & Howe, J. T. B. (1999). Corporate risk management for multinational corporations: Financial and operational hedging policies. *European Finance Review, 2,* 229–246.

Cronin, B., & Ding, Y. (2011). Popular and/or prestigious? Measures of scholarly esteem. *Information Processing Management, 47,* 80–96.

Dada, M., & Hu, Q. (2008). Financing newsvendor inventory. *Operations Research Letters, 36,* 569–573.

Ding, Q., Dong, L., & Kouvelis, P. (2007). On the integration of production and financial hedging decisions in global markets. *Operations Research, 55,* 470–489.

Ding, Y., Yan, E., Frazho, A., & Caverlee, J. (2009). PageRank for ranking authors in co-citation networks. *Journal of the American Social Information Science and Technology, 60,* 2229–2243.

Fahimnia, B., Tang, C. S., Davarzani, H., & Sarkis, J. (2015). Quantitative models for managing supply chain risks: A review. *European Journal of Operational Research, 247,* 1–15.

Farris, M. T., II, & Hutchison, P. D. (2002). Cash-to-cash: The new supply chain management metric. *International Journal of Physical Distribution & Logistics Management, 32,* 288–298.

Fisman, R., & Love, I. (2003). Trade credit, financial intermediary development and industry growth. *The Journal of Finance, 58,* 353–374.

Froot, K. A., Scharfstein, D. S., & Stein, J. C. (1993). Risk management: Coordinating corporate investment and financing policies. *The Journal of Finance, 48,* 1629–1658.

Fullerton, R., McWatters, C., & Fawson, C. (2003). An examination of the relationships between JIT and financial performance. *Journal of Operations Management, 21,* 383–404.

Gamba, A., & Triantis, A. J. (2014). Corporate risk management: Integrating liquidity, hedging, and operating policies. *Management Science, 60,* 246–264.

Gaur, V., & Seshadri, S. (2005). Hedging inventory risk through market instruments. *Manufacturing & Service Operations Management, 7,* 103–120.

Gelsomino, L. M., Mangiaracina, R., Perego, A., & Tumino, A. (2016). Supply chain finance: A literature review. *International Journal of Physical Distribution & Logistics Management, 46,* 348–366.

Grosse-Ruyken, P. T., Wagner, S. M., & Jönke, R. (2011). What is the right cash conversion cycle for your supply chain? *International Journal of Services and Operations Management, 10,* 13–29.

Guillen, G., Badell, M., & Puigjaner, L. (2007). A holistic framework for short-term supply chain management integrating production and corporate financial planning. *International Journal of Production Economics, 106,* 288–306.

Gupta, S., & Dutta, K. (2011). Modeling of financial supply chain. *European Journal of Operational Research, 211,* 47–56.

Hankins, K. W. (2011). How do financial firms manage risk? Unraveling the interaction of financial and operational hedging. *Management Science, 57,* 2197–2212.

Hendricks, K. B., & Singhal, V. R. (2003). The effect of supply chain glitches on shareholder wealth. *Journal of Operations Management, 21,* 501–523.

Hendricks, K. B., & Singhal, V. R. (2005). An empirical analysis of the effect of supply chain disruptions on long-run stock price and equity risk of the firm. *Production and Operations Management, 14,* 35–52.

Hofmann, E. (2005). Supply chain finance: Some conceptual insights. In R. Lasch & C. G. Janker (Eds.), *Logistik management—Innovative Logistikkonzepte* (pp. 203–214). Wiesbaden: Deutscher Universitätsverlag.

Hofmann, E., & Kotzab, H. (2010). A supply chain-oriented approach of working capital management. *Journal of Business Logistics, 31,* 305–330.

Hommel, U. (2003). Financial versus operative hedging of currency risk. *Global Finance Journal, 14,* 1–18.
Houston, J. F., Lin, C., & Zhu, Z. (2016). The financial implications of supply chain changes. *Management Science, 62,* 2520–2542.
Huang, Y. F., & Hsu, K. H. (2008). An EOQ model under retailer partial trade credit policy in supply chain. *International Journal of Production Economics, 112,* 655–664.
Huchzermeier, A., & Cohen, M. A. (1996). Valuing operational flexibility under exchange rate risk. *Operations Research, 44,* 100–113.
Huo, B., Han, Z., Zhao, X., Zhou, H., Wood, C. H., & Zhai, X. (2013). The impact of institutional pressures on supplier integration and financial performance: evidence from China. *International Journal of Production Economics, 146,* 82–94.
Iacono, U. D., Reindorp, M., & Dellaert, N. (2015). Market adoption of reverse factoring. *International Journal Physical Distribution & Logistics Management, 45,* 286–308.
Kazaz, B., Dada, M., & Moskowitz, H. (2005). Global production planning under exchange-rate uncertainty. *Management Science, 51,* 1101–1119.
Kim, Y., Mathur, I., & Nam, J. (2006). Is operational hedging a substitute for or a complement to financial hedging? *Journal of Corporate Finance, 12,* 834–853.
Klapper, L. (2006). The role of factoring for financing small and medium enterprises. *Journal of Banking & Finance, 30,* 3111–3130.
Koberstein, A., Lukas, E., & Naumann, M. (2013). Integrated strategic planning of global production networks and financial hedging under uncertain demand and exchange rates. *Business Research, 6,* 215–240.
Kogut, B., & Kulatilaka, N. (1994). Operating flexibility, global manufacturing, and the option value of a multinational network. *Management Science, 40,* 123–139.
Kouvelis, P., Dong, L., Boyabatli, O., & Li, R. (2012). Integrated risk management: A conceptual framework with research overview and applications in practice. P. Kouvelis, L. Dong, O. Boyabatli & R. Li (Eds.), *Handbook of integrated risk management in global supply chains* (pp. 3–12). New York: Wiley.
Kouvelis, P., Li, R., & Qing, D. (2013). Managing storable commodity risks: The role of inventories and financial hedge. *Manufacturing & Service Operations Management, 15,* 507–521.
Kouvelis, P., & Zhao, W. (2012). Financing the newsvendor: Supplier vs. bank, and the structure of optimal trade credit contracts. *Operations Research, 60,* 566–580.
Kouvelis, P., & Zhao, W. (2016). Supply chain contract design under financial constraints and bankruptcy costs. *Management Science, 62,* 2341–2357.
Lanier, D., Jr., Wempe, W. F., & Zacharia, Z. G. (2010). Concentrated supply chain membership and financial performance: Chain- and firm-level perspectives. *Journal of Operations Management, 28,* 1–16.
Lee, C. H., & Rhee, B. D. (2010). Coordination contracts in the presence of positive inventory financing costs. *International Journal of Production Economics, 124,* 331–339.
Longinidis, P., & Georgiadis, M. C. (2011). Integration of financial statement analysis in the optimal design of supply chain networks under demand uncertainty. *International Journal of Production Economics, 129,* 262–276.
Longinidis, P., & Georgiadis, M. C. (2014). Integration of sale and leaseback in the optimal design of supply chain networks. *Omega, 47,* 73–89.
Mello, A. S., Parsons, J. E., & Triantis, A. J. (1995). An integrated model of multinational flexibility and financial hedging. *Journal of International Economics, 39,* 27–52.
Meulbroek, L. (2002). The promise and challenge of integrated risk management. *Risk Management and Insurance Review, 5,* 55–66.
Miller, K. (1992). A framework for integrated risk management in international business. *Journal of International Business Studies, 23,* 311–331.
Modigliani, F., & Miller, M. (1958). The cost of capital, corporation finance, and the theory of investment. *The American Economic Review, 48,* 261–297.
Myers, S. C. (1984). The capital structure puzzle. *The Journal of Finance, 39,* 575–592.

Park, J. H., Kazaz, B., & Webster, S. (2017). Risk mitigation of production hedging. *Production and Operations Management, 26,* 1299–1314.

Pfohl, H.-C., & Gomm, M. (2009). Supply chain finance—Optimizing financial flows in supply chains. *Logistics Research, 1,* 149–161.

Protopappa-Sieke, M., & Seifert, R. W. (2010). Interrelating operational and financial performance measurements in inventory control. *European Journal of Operational Research, 204* 439–448.

Raghavan, N. S., & Mishra, V. K. (2011). Short-term financing in a cash-constrained supply chain. *International Journal of Production Economics, 134,* 407–412.

Randall, W., & Farris, M., II. (2009). Supply chain financing: Using cash-to-cash variables to strengthen the supply chain. *International Journal of Physical Distribution & Logistics Management, 39,* 669–689.

Secomandi, N. (2010a). Optimal commodity trading with a capacitated storage asset. *Management Science, 56,* 449–467.

Secomandi, N. (2010b). On the pricing of natural gas pipeline capacity. *Manufacturing & Service Operations Management, 12,* 393–408.

Secomandi, N., & Wang, M. X. (2012). A computational approach to the real option management of network contracts for natural gas pipeline transport capacity. *Manufacturing & Service Operations Management, 14,* 441–454.

Seifert, D., Seifert, R. W., & Protopappa-Sieke, M. (2013). A review of trade credit literature: Opportunities for research in operations. *European Journal of Operational Research, 231,* 245–256.

Simchi-Levi., D. (2014). OM research: From problem-driven to data-driven research. *Manufacturing & Service Operations Management, 16,* 2–10.

Tang, C. S. (2017). Identifying research topics in operations management. *Manufacturing & Service Operations Management, 19,* 338–346.

Tang, O., & Musa, N. S. (2011). Identifying risk issues and research advancements in supply chain risk management. *International Journal of Production Economics, 133,* 25–34.

Taylor, T. A., & Xiao, W. (2014). Subsidizing the distribution channel: Donor funding to improve the availability of malaria drugs. *Management Science, 60,* 2461–2477.

Triantis, A. J. (2000). Corporate risk management: Real options and financial hedging. *Journal of Applied Corporate Finance, 13,* 64–73.

Trigeorgis, L. (1993). Real options and interactions with financial flexibility. *Financial Management, 22,* 202–224.

Turcic, D., Kouvelis, P., & Bolandifar, E. (2015). Hedging commodity procurement in a bilateral supply chain. *Manufacturing & Service Operations Management, 17,* 221–235.

van Mieghem, J. A. (2007). Risk mitigation in newsvendor networks: Resource diversification, flexibility, sharing, and hedging. *Management Science, 53,* 1269–1288.

van Mieghem, J. A. (2012). Risk management and operational hedging: An overview. P. Kouvelis, L. Dong, O. Boyabatli & R. Li (Eds.), *Handbook of integrated risk management in global supply chains* (pp. 13–49). New York: Wiley.

Vickery, S. K., Jayaram, J., Droge, C., & Calantone, R. (2003). The effects of an integrative supply chain strategy on customer service and financial performance: An analysis of direct versus indirect relationships. *Journal of Operations Management, 21,* 523–539.

van der Vliet, K., Reindorp, M. J., & Fransoo, J. C. (2015). The price of reverse factoring: Financing rates vs. payment delays. *European Journal of Operational Research, 242,* 842–853.

Wang, D., Song, C., & Barabási, A.-L. (2013). Quantifying long-term scientific impact. *Science, 342,* 127–132.

Wang, Y., Gilland, W., & Tomlin, B. (2010). Mitigating supply risk: Dual sourcing or process improvement? *Manufacturing & Service Operations Management, 12,* 489–510.

Wuttke, D. A., Blome, C., Foerstl, K., & Henke, M. (2013a). Managing the innovation adoption of supply chain finance—Empirical evidence from six European case studies. *Journal of Business Logistics, 34,* 148–166.

Wuttke, D. A., Blome, C., & Henke, M. (2013b). Focusing the financial flow of supply chains: An empirical investigation of financial supply chain management. *International Journal of Production Economics, 145,* 773–789.

Xiao, S., Sethi, S. P., Liu, M., & Ma, S. (2017). Coordinating contracts for a financially constrained supply chain. *Omega, 72,* 71–86.

Zhao, L., & Huchzermeier, A. (2015). Operations-finance interface models: A literature review and framework. *European Journal of Operational Research, 244,* 905–917.

Uncited References

Blome, C., & Schoenherr, T. (2011). Supply chain risk management in financial crises—a multiple case-study approach. *International Journal of Production Economics, 134,* 43–57.

Cachon, G., & Terwiesch, C. (2011). *Matching supply with demand: An introduction to operations management.* McGraw-Hill/Irwin.

Carter, C. R., & Rogers, D. S. (2008). A framework of sustainable supply chain management: Moving toward new theory. *International Journal of Physical Distribution & Logistics Management, 38,* 360–387.

Cohen, M. A., & Huchzermeier, A. (1999). Global supply chain management: A survey of research and applications. S. Tayur, R. Ganeshan & M Magazine (Eds.), *Quantitative models for supply chain management.* Norwell, MA: Kluwer Academic Publishers.

Corsten, D., & Saraf, B. (2008). *Nestle Russia LLC: Supplier finance programme.* Madrid: Instituto de Empressa business school case.

Cosh, A., Cumming, D., & Hughes, A. (2009). Outside entrepreneurial capital. *The Economic Journal, 119,* 1494–1533.

Cox, L. A., Jr. (2008). What's wrong with risk matrices? *Risk Analysis, 28,* 497–512.

Dasu, S., & Li, L. (1997). Optimal operating policies in the presence of exchange rate variability. *Management Science, 43,* 705–722.

Fullerton, R., McWatters, C., & Fawson, C. (2003b). An examination of the relationships between JIT and financial performance. *Journal of Operations Management, 21,* 383–404.

Gupta, R. K., Bhunia, A. K., & Goyal, S. K. (2009). An application of Genetic Algorithm in solving an inventory model with advance payment and interval valued inventory costs. *Mathematical Computer Modelling, 49,* 893–905.

International Integrated Reporting Committee [IIRC]. (2011). Towards integrated reporting—Communicating value in the 21st century. International Integrated Reporting Committee.

Kouvelis, P., & Zhao, W. (2016b). Supply chain contract design under financial constraints and bankruptcy costs. *Management Science, 62,* 2341–2357.

Lai, G., Debo, L. G., & Sycara, K. (2009). Sharing inventory risk in supply chain: The implication of financial constraint. *Omega, 37,* 811–825.

Lee, H. L., Padmanabhan, V., & Whang, S. (1997). Information distortion in a supply chain: The bullwhip effect. *Management Science, 43,* 546–558.

Lessard, D., & Nohria, N. (1990). Rediscovering functions in the MNC: The role of expertise in firms' response to shifting exchange rates. In C. Bartlett, Y. Doz & G. Hedlund (Eds.), *Managing the global firm* (pp. 186–212). London: Routledge.

Lessard, D. R., & Zaheer, S. (1996). Breaking the silos: Distributed knowledge and strategic responses to volatile exchange rates. *Strategic Management Journal, 17,* 513–533.

Mauer, D. C., & Triantis, A. J. (1994). Interactions of corporate financing and investment decisions: A dynamic framework. *The Journal of Finance, 49,* 1253–1277.

Mello, A. S., & Parsons, J. E. (2000). Hedging and liquidity. *The Review of Financial Studies, 13,* 127–153.

Rajan, R. G., & Zingales, L. (1998). Financial dependence and growth. *American Economic Review, 88,* 559–586.

Ritchie, B., & Brindley, C. (2007). An emergent framework for supply chain risk management and performance measurement. *Journal of the Operational Research Society, 58,* 1398–1411.

Sting, F. J., & Huchzermeier, A. (2012). Dual sourcing: Responsive hedging against correlated supply and demand uncertainty. *Naval Research Logistics, 59,* 69–89.

Tang, C. S., & Tomlin, B. (2008). The power of flexibility for mitigating supply chain risks. *International Journal of Production Economics, 116,* 12–27.

Tang. C.S., Wu, J., & Yang, S.A. (2017). Sourcing from suppliers with financial constraints and performance risk. *Manufacturing & Service Operations Management,* forthcoming.

Thangam, A. (2012). Optimal price discounting and lot-sizing policies for perishable items in a supply chain under advance payment scheme and two-echelon trade credits. *International Journal of Production Economics, 139,* 459–472.

Tripp, M. H., Haria, C. S., Hilary, N., Morgan, K., Orros, G. C., Perry, G. R., & Tahir-Thomson, K. (2008). *Enterprise risk management from the general insurance actuarial perspective.* Paper presented to the Institute of Actuaries.

Wandfluh, M., Hofmann, E., & Schoensleben, P. (2016). Financing buyer–supplier dyads: An empirical analysis on financial collaboration in the supply chain. *International Journal of Logistics Research and Applications, 19,* 1–16.

Zhao, L., & Huchzermeier, A. (2017). Integrated operational and financial hedging with capacity reshoring. *European Journal of Operational Research, 260,* 557–570.

The manufacturer's authorised representative in the EU is Springer Nature Customer Service Centre GmbH, Europaplatz 3, 69115 Heidelberg, Germany. If you have any concerns regarding our products, please contact ProductSafety@springernature.com

Printed and bound by CPI Group (UK) Ltd, Croydon, CR0 4YY

30/09/2026

02225658-0002